VALIANT BOYS

PREVIOUS BOOKS BY TONY BLACKMAN

Flight Testing to Win (Autobiography paperback)
ISBN 978-0-9553856-4-3, 0-9553856-4-4
Published Blackman Associates 2005

Vulcan Test Pilot
ISBN 978-1-906502-30-0
Published Grub Street 2007

Tony Blackman Test Pilot
ISBN 978-1-906502-28-7
Published Grub Street 2009

Vulcan Owner's Workshop Manual
ISBN 978-1-844258-31-4
Published Haynes 2010

Nimrod Rise and Fall
ISBN 978-1-908117-79-3
Published by Grub Street 2011

Victor Boys
ISBN 978-1-908117-45-8
Published Grub Street 2012

Vulcan Boys
ISBN 978-1-909808-08-9
Published Grub Street 2014

FICTION

A Flight Too Far
ISBN 978-0-9553856-3-6, 0-9553856-3-6
Published Blackman Associates

The Final Flight
ISBN 978-0-9553856-0-5, 0-9553856-0-1
Published Blackman Associates

The Right Choice
ISBN 978-0-9553856-2-9, 0-9553856-2-8
Published Blackman Associates

Flight to St Antony
ISBN 978-0-9553856-6-7 0-9553856-6-0
Published Blackman Associates

Now You See It
ISBN 978-0-9553856-7-4, 0-9553856-7-9
Published Blackman Associates

Dire Strait
ISBN 978-0-9553856-8-1
Published Blackman Associates

VALIANT BOYS

True Stories from the Operators of
the UK's First Four-Jet Bomber

TONY BLACKMAN
ANTHONY WRIGHT

Grub Street · London

Published by
Grub Street
4 Rainham Close
London
SW11 6SS

Copyright © Grub Street 2014
Copyright text © Tony Blackman 2014, © Anthony Wright 2014

A CIP record for this title is available from the British Library

ISBN-13: 978-1-909808-21-8

Cover design by Sarah Baldwin

Printed and bound by Finidr

Grub Street Publishing only uses
FSC (Forest Stewardship Council) paper for its books.

CONTENTS

DEDICATION

To the many Valiant aircrews and ground crews, some sadly
no longer with us, who operated the aircraft from 1952 to 1964
helping to guard the United Kingdom through many
troubled times.

FOREWORD

I was very pleased when Tony Blackman asked me to write a foreword for the book *Valiant Boys*. It is good to have a book written by aircrew and ground crew telling their stories and how they operated the aircraft so that all these things are recorded and not forgotten. The Valiant was the first of the V bombers and I know that a lot of these great people who flew and maintained the aircraft feel that what they did and their aircraft itself gets forgotten and overshadowed by the Vulcan and the Victor but that's not my view at all. The Valiant did exactly what was required at the time to help defend the UK and the Western World. It helped to develop the UK's atom and hydrogen bombs and was the only one of the V bombers that actually dropped a live weapon. I was fortunate enough actually to see these terrible weapons explode and it is partly thanks to the Valiant initially providing the UK's nuclear deterrent in the mid 'fifties that we never had to use the weapons in anger.

I was exceptionally lucky during my career in the RAF not only to have been involved in bringing all three V bombers into service but also to have had the opportunity to command 214 Valiant Squadron in 1959 and 1960 while we were developing flight refuelling. It was a real thrill to demonstrate the effectiveness of the aircraft by being refuelled over Kano by another Valiant of the squadron and breaking the non-stop record from the UK to Cape Town.

Tony was an Avro test pilot but he flew and tested all three V bombers. He has already written *Vulcan Boys* and *Victor Boys* but he wanted to write *Valiant Boys* before it was too late because he knows what a wonderful aircraft the Valiant was and because he is determined that it shall not be forgotten. The Valiant did everything it was asked to do, not only as a deterrent but as a strategic reconnaissance aircraft; in addition we must always remember that the Vulcan and the Victor flying thousands of miles across the South Atlantic in the Falklands campaign with its incredible reliance on flight refuelling, could not have taken place without the pioneering work of the Valiant.

We are very lucky in this country to have been able to design and build three superb four-jet bombers when jet aircraft and the jet engine were still in their infancy; with the V bombers we helped to preserve the world from nuclear conflict. Looking back we must not forget that the development of the engines to power the V bombers enabled Rolls-Royce to continue improving their expertise so that their engines now power civil and military aircraft all round the world to the benefit of the UK. Long may the memory of the Valiant, Victor and Vulcan continue.

Marshal of the Royal Air Force
Sir Michael Beetham GCB, CBE,
DFC, AFC, FRAeS
July 2014

ACKNOWLEDGEMENTS

As I finished writing *Vulcan Boys* I decided it would be a good idea to complete the trio of the V bombers. The Valiant is not as well known as it should be despite being the UK's first four-jet bomber beating the Vulcan by fifteen months. However, though I did a little testing of the aircraft and knew the firm's test pilots, I did not know any operators which presented me with a real problem. Luckily in putting together *Vulcan Boys* I had got to know Anthony Wright who helped me enormously not only in writing a chapter but also in checking the material for mistakes. I was really delighted therefore, when he agreed to co-author *Valiant Boys*; he has written two chapters relating his own experiences of the Valiant but, just as important, he seemed to know all the Valiant people and persuaded many of them to contribute to the book.

We are of course very grateful to Sir Michael Beetham for agreeing to write a foreword to the book; it will give a lot of pleasure to the operators and their families to know that all their work is appreciated and remembered.

It has been quite difficult to find people who operated the aircraft in its early days and so I was forced to do some research myself about the atomic weapon development. Luckily again Don Briggs, World War One flight engineer and later Valiant co-pilot who was at Christmas Island, had dictated his experiences and Iris Jenkins, my erstwhile next-door neighbour, very kindly transcribed them. In addition Bill Evans was a corporal airframe technician at both Maralinga and Christmas Island and kept a real diary. Finally Clive Cox, an airframe mechanic and David Kent, a wireless technician, were at Christmas Island to add their impressions.

For the other chapters Anthony and I are grateful to Shaun Broaders, Milt Cottee, Gordon Dyer, John Foot, Terry Gladwell, Graeme Kerr, Alan McDonald, Roy Monk, Robby Robinson, Russ Rumbol, David Sykes, Peter West and to the Express Newspapers/Express Syndication for permission to use the prologue; also for the short stories we were delighted to receive from John Foster, Brian Loveday, Eric Macey, Bryan Montgomery, John Saxon, Peter Sharp, Keith Walker and Brian Yates. We must apologise for anyone we have unintentionally omitted. For Chapters Eighteen and Nineteen of the book we are grateful to Richard Harris, whose father was killed in Valiant accident XD864, for all the investigation he did on the metal used in the Valiant spar.

David Wright's contribution to the book was three really splendid paintings of the Valiant in action but he also flew Valiants for six years on 214 Squadron and then the Operational Conversion Unit (OCU) including two bombing raids on Suez.

Re photographs, hopefully we have acknowledged all the ones that we have received though some are in the public domain and we apologise if we have used any that we have not acknowledged. We would particularly like to thank Nick Stroud and his splendid *Aviation Historian* for his contributions. Also Martyn

Chorlton of Old Forge Publications for letting us use some photos from Robby Robinson's *Jet Bomber Pilot: Autobiography of a V-bomber Pilot*, Peter Jacobs and Frontline Publishers for the use of Sir Michael Beetham's photograph, *The Daily Telegraph* and lastly the RAF Museum for flight deck pictures of XD818 which dropped the UK's first H bomb.

Finally, we would like to thank our publisher for all his support, with his suggestions, superb editing and the speed in which he has turned the draft into a book to be proud of.

Tony Blackman and Anthony Wright
July 2014

PREFACE

This book is about the first of the three V bombers which were built to guard the United Kingdom during the Cold War. The object in producing it is to put on record all the things the aircraft did with memories written by the operators themselves. First-hand accounts and personal opinions give an authenticity that is impossible to match just relating conversations.

One cannot help but be impressed reading of the long hours spent by the crews, ground as well as flight, sitting by the aircraft at the readiness platforms waiting to be scrambled to start a third world war. Thankfully it never happened but might well have done if the crews had not been sitting there, ready. The tedium was relieved by lone rangers flying to the United States and to the Mediterranean and the Far East.

The aircraft had another very important strategic reconnaissance role which was also put to good use in a civilian application for detailed large-scale mapping. The Valiant also developed flight refuelling enabling support of the RAF fighters patrolling the UK, helping the fighters to stage to overseas bases and later for the Victor K2s to support the Falklands campaign.

Thanks again to all the people who helped write this book, and for making it such a compulsive read and for making sure the Valiant's importance as the first V bomber is not forgotten.

NB: Though the book has been put together and written jointly by Anthony Wright and myself, all the editorial comments in the book which are in the first person are written by myself, Tony Blackman. The reader will also note that there are two styles to the text layout in the book. Indented and full out. All indented text is my commentary.

PROLOGUE

THE FIRST PROTOTYPE
Jock Bryce

The first prototype Vickers Valiant. (*via www.aviastar.org*)

Mutt Summers, chief test pilot of Vickers Armstrong was taxiing the prototype Vickers Valiant down the runway at Wisley, Weybridge. He was taxiing, as I thought, far too fast, jamming the brakes on fiercely, screeching round corners, throwing the aircraft about, almost trying to break something. "Mutt, why do you have to abuse this airplane?" It seemed an extraordinary way to treat a brand new prototype aeroplane, one that we had flown for the first time, the only one of its kind in existence and I said so.

"Look Jock," said Summers, "if I can't break this aeroplane then neither can the customers." To be a test pilot you had to flog the aircraft to its limits and beyond; that was the lesson Mutt was trying to get across.

It was shortly after this that Mutt Summers retired and I was appointed to succeed him. My first task was to push the Valiant, the first of the V bombers, to its limits and beyond in the air. The aircraft, with its four Rolls-Royce Avon turbojet engines and its swept back wings and tail surfaces, was at that time on the secret list; it was the front runner of the V bomber force that had recently been announced, superior in performance and striking power to any other military aircraft in the world. It first flew on 18th May 1951 and the public had seen it only once when I had flown it at the Farnborough Air Show the previous year.

One Saturday afternoon – it was 11th January 1952 – after nearly 100 hours of test flying directed at proving the airplane for its role of high speed, high altitude, heavy bomber we took off in the prototype aircraft WB210 from Hurn airport near Bournemouth where we were based on a rather unusual test flight, the first part of which had nothing to do with the Valiant test programme. The object was to measure and estimate the noise levels likely to be experienced in a four-jet transport aircraft — the V 1000 — which was already in the project design stage. The tests involved flying the Valiant at varying engine settings with the two inboard engines stopped.

I had a full crew on board; Brian Foster, my second pilot was an RAF squadron leader who had been sent to Vickers Armstrong from Bomber Command as their project pilot on the Valiant. He sat next to me on the flight deck. Behind and below us was the crew compartment which stretched the width of the fuselage; the technical crew consisted of three men – Roy Holland, John Prothero-Thomas and Jeffrey Montgomery. We flew in the direction of Wisley for no reason other than this was an area we knew. It was a fine afternoon and we were turning over the river near Kingston-on-Thames with the two inboard engines stopped when we had completed the special tests we had been detailed to do. I carried out the procedure for restarting the engines but the starboard engine did not relight. These engines, like the airplane, were in the proving stage but I was not unduly perturbed. There was no reaction when I tried a second time but it lit up all right on the third attempt.

We had been airborne for only about half an hour so we had enough fuel to continue with our normal development programme. A few weeks earlier I had opened the bomb doors for the first time and due to a structural failure one of the doors had fallen off, landing in the New Forest. So I turned away from the built-up area around Kingston and headed south-west for the coast.

In the next few minutes we did a series of checks on the bomb doors noting the effect of trim buffet and drag by which time we were over Portsmouth. Suddenly, as I was opening the bomb door for the third time I noticed that some of the instruments on the starboard side were not registering. All these instruments were duplicated in the rear crew compartment and I called up Roy Holland.

"How are you doing on the starboard engine, Roy?"

"That's funny. We've lost a lot of the engine instruments on that side."

The engines were still running and there was no other indication that anything was wrong and I headed west at once towards Hurn which was only about three minutes flying away. I called them up on the voice radio.

"I'm returning early."

That would alert them on the ground and I knew they would give me priority as I was flying a prototype airplane. I was still in full control and experiencing no difficulty but I was trying hard to analyse what was going on. I decided it must be some electrical fault and I thought perhaps we had lost one of the bomb doors again. So I got the crew to check but everything there was in order.

We were crossing Southampton Water on our way to Hurn when there was a loud thump in the airplane, a dull muffled explosion which I heard clearly in spite of the helmet and oxygen equipment I was wearing.

"Any of you fellows hear a bump in the airplane?"

"Yes, we did."

I couldn't imagine what it might have been and so instead of descending so as to cross Hurn at 2,000ft preparatory to getting into the circuit for landing, I sat there at 7,000ft trying to work out what was wrong. The engine instruments that had failed still showed no signs of life but the fault could still be an electrical one.

We were right over the top of the airfield at Hurn when I suddenly realised that I had no lateral control. I moved the control column from port to starboard and there was no response. It was a terrifying sensation.

These were power controls but the control column itself felt normal enough. The trouble was it no longer seemed as if it was connected to anything. I did not know it but right across Southampton Water and along the coastal strip from Calshot to Christchurch the Valiant had been spilling a trail of molten wreckage. Unknown to me the interior of the starboard wing was on fire. The push-pull rod controlling the ailerons had burnt right through.

The strength of the main spar on the side was being reduced by the heat and as it weakened it sought to relieve the load on itself by twisting and warping. This was giving me a different angle of incidence on one wing from the other. The aircraft began to force its way into a turn to the left and I could not hold it.

Coming up below me was the hinterland of Bournemouth with the coastline to my left. I was well beyond Hurn airport. I called my crew: "I've lost my aileron control. Have a look through the periscope and see if you can see any structural fault. That thump must have been caused by something."

By this time the aircraft was flying along in a steep bank and I was using rudder to bring me around in a left-hand circuit behind Bournemouth and Poole and slightly out to sea towards Swanage, trying all the time to position myself to come back to Hurn. As we crossed the coast on this wide and uncertain circuit, the aircraft began to rock violently. For a time it tipped over to a 50° bank and I couldn't get it up. Lacking aileron control I knew I would never be able to get the airplane down at Hurn. Instinctively I decided to make for Boscombe Down on Salisbury Plain where there was a much longer runway. I called Hurn on the radio. "I am declaring an emergency. I have no aileron control. I am going to attempt a landing at Boscombe Down."

I managed to bring the plane round in a complete circle, crossing the coast again at Christchurch and heading inland to Boscombe Down. I called Boscombe and asked them to clear the circuit for me, but even as I did so I knew in my heart of hearts that I didn't have sufficient control to get down even on that huge runway. Yet I still could not convince myself that for the first time in my life I was going to tell my crew to bale out.

The aircraft was still swinging violently from port to starboard, though I was able to maintain height. I forced myself to face the fact that the bale-out order would have to be given. The plane was rapidly running away from me. I called the crew. "We're in big trouble here. I can't control this airplane."

Brian Foster, my second pilot, was looking back over his right shoulder towards the starboard wing. From where I sat I couldn't see the wing on that side but Brian could see the outer half. "Gee, it's on fire."

If the outer part of the wing was on fire then the inner half where the engines were and where the fire must have started must be an inferno. "It's coming out of the leading edge of the wing," called Brian, "smoke and flame." We would have to get out quickly. Once again I called the crew, "Abandon the aircraft, bale out, bale out."

I held the airplane for the next few seconds on the elevators and by pedalling the rudder but we were charging through the sky almost completely out of control. On my bale-out order the first thing the men behind me had to do was to disconnect their headsets so I heard nothing from them.

Brian and I blew off the hood of the pilots' compartment and then I waited for Brian to go. He was looking back over his left shoulder, peering back into the crew compartment behind us waiting to see the last of the technical crew bale out. They had no ejector seats but would bale out in the old-fashioned way through the port side door of the aircraft.

I knew if we didn't get out soon it would be too late and I turned and punched Brian on the arm, "Come on Brian, get out." The next moment I smelt a cordite charge which was his seat ejecting as he pulled the blind down in front of his face. Almost simultaneously, just as his seat ascended there was an explosion behind us as the starboard wing blew off. The fuel on that side had ignited and 3,000 gallons of kerosene had gone up.

The explosion threw the Valiant right over on its side. I was still in the cockpit. I was using one of the early Martin Baker ejector seats. I was only the third test pilot to use one and Brian Foster had only just beaten me to it. Unlike the modern versions these early seats did not operate fully automatically so I had to do some of the drill manually. There were three stages to this in strict sequence. The first was to eject the seat with me safely in it. The second to get myself clear of the seat and the third to open the parachute. It occurred to me I would only have one chance to get all this right.

I disconnected the oxygen supply to my mask, then realised there was no need to do this because this was one part of the ejection sequence that was automatic. I then reconnected the mask wasting precious seconds vaguely conscious that what I was doing was unnecessary but anxious to get the whole procedure right.

I went through the drill carefully in my mind and then, bracing myself for the explosive force that was about to lift me vertically through the hole in the roof of the cockpit, I reached above my head for the lever and pulled the blind down in front of my face.

I must have blacked out as I ejected because the next thing I knew above me was a brilliant blue sky. Below me lay a brown and green England. Then it was the ground that was above me and the sky below, the whole panorama interchanging in rapid succession as I somersaulted through the sky. Again and again I saw the horizon rising at me as I toppled over and over.

Around me, as though in a vacuum, huge jagged chunks of metal cartwheeled with me, maintained their position. The aeroplane must have broken up as I went. I had no realisation of being blasted out and had no idea of what had happened to the others. The only thought in my mind was for myself. I was still strapped in my seat and paralysed by the centrifugal force of the somersaults. It was going to be a criminal way to die.

The horizon was coming round a little less violently each time and I realised that the gyrations were slowing down. When they finally stopped I found myself sitting in the ejector seat at an unknown altitude but one that I knew must give me very little time if I was to get down safely. It was an extraordinary sensation, positioned in a sitting situation, plunging through the sky.

I began to think about getting out of the seat. My movements felt strangely unnatural, as if the force of gravity had for the moment been suspended. I did not know how I was going to free myself. The trouble was that I had nothing to push against, no floor to press my feet on and lift myself out.

Two sets of straps encased my body, one set strapping me into the seat and the other strapping me to the parachute. Which buckles must I disconnect first? Here was another vital choice and I wasted more valuable time tracing the buckles to their source. The parachute buckles were white, the seat buckles were blue. I repeated it carefully. Seat buckles blue. Had I got it right? I was terrified in case I should unfasten the wrong straps although the distinction was obvious and in calmer moments I knew it well. I hesitated a moment longer, loath to make the final decision, then unlocked the blue buckle.

I had expected the straps to fly open and the seat to fall away, but nothing happened. Was there some other fastening? I couldn't find one. Somehow I would have to throw the buckle off and climb out of the seat.

The parachute fitted into a tight little bin in the seat and I had to separate the two before I pulled the ripcord. Otherwise the seat might foul the canopy. I kept trying to push the seat away from me, systematically at first and then with the strength of desperation but I still couldn't find some means of levering myself out. Now that the seat was stabilised I was able to take a look round to see where I was. I could see the coastline a few miles ahead and I knew I was coming down well inland. Suddenly the ground seemed frighteningly close. I tried once more to squirm out of my seat, then gave up and pulled my ripcord.

For a second time I passed from one state into another without any realisation of what I had done. The canopy of the parachute was suddenly above me, fully developed but I had felt no shock as it opened. The ejector seat had gone. The rigging lines of the parachute were hopelessly entwined, and I was pirouetting round in one direction, over swinging until the motion almost stopped and then whirling back in the other. I had to stop this fierce rotation before I hit the ground.

I grasped the rigging lines above my head and as I did so I hit something. My feet were caught in a tree. The effect of hitting the tree was to turn me over on to my back and I finished up by landing horizontally in a field. I hit the ground heavily and lay on my back completely winded. The events of the last few minutes had vanished from my mind and I could not recall them. I could not think what had happened. Even when I recovered my breath I still lay on my back, relaxing, enjoying these few precious moments of amnesia. I thought I had probably broken my back and I was not yet prepared to make the experiments that would confirm or disprove it.

The first person I saw was a woman in a kitchen apron peering down at me. She looked about thirty feet high. She asked the question that all English women ask when a parachute falls into their back garden. "Would you like a cup of tea?" I nodded and she left, reappearing shortly afterwards with the tea. I still didn't understand what had happened to me.

The first thing I realised was that of all the vast areas of Southern England which must have been completely unknown to me I had landed at one of the few spots I knew. I actually recognised the Nissen huts of No 8 Site at Holmsley South airfield in the New Forest where I had been stationed during the war. I could see a black column of smoke about 800 yards away, but for several minutes I was still hazy about what had happened, though I began to wonder vaguely about my crew.

I was now surrounded by a little knot of people. Suddenly the events of the last thirty minutes crowded in on me. My back was painful but it seemed all right, so I asked them to get me to a telephone. There was a phone box outside No 8 Site. I remembered it and got through to my number two, Brian Trubshaw who was on the ground at Hurn waiting for news and I told him we had lost the aircraft. An ambulance man came along and treated my face which had evidently been torn when I hit the tree. Then a police car drove up and I got in. They were searching for the rest of the crew.

Two of the technicians had fallen quite near me and we soon found them, happily unhurt. The third technician Roy Holland had landed in a tree a long way to the south and when we got there the local fire brigade were trying to get him down. When they finally rescued him he was blue with cold.

Meanwhile I had been stunned by the discovery that Brian Foster, my second pilot, had been killed. He had ejected before me and at first I was completely incredulous until I had time to think it out. What I believed happened was that the starboard wing exploded as he went cushioning the upward force of the ejection mechanism so that he failed to clear the tail. The airplane then slipped into a violent bank to port and when I ejected a moment later I was clear of any obstruction.

When we had rounded up the rest of the crew we drove back to the flight test office at Hurn. It was then my task to phone George Edwards who by this time was managing director of Vickers and tell him we'd lost the Valiant prototype.

"What do you mean, you've lost it?"

"We baled out and the aircraft's a write off."

"Is everyone all right?"

"Unfortunately not, Brian Foster is killed."

"I'll be down as soon as I can."

By the time George Edwards arrived at Hurn I and my crew had dictated to my secretary an exact report of the incident as we individually remembered it, before our judgement could be swayed by additional knowledge or hindsight and while our minds were clear. We spent the rest of the evening with George Edwards going over what had happened in great detail and then, on Sunday morning, we went out to look at the wreckage.

As one would expect the aircraft had broken up completely. The port wing was gone, the starboard wing was burnt out and so was the fuselage. In a cavity in the ground, about four feet down, and three times the width of the flight deck was the spot where I had been sitting. I turned my back on the wreckage and walked away, watching the others from a distance. It was easy to imagine myself still in that cockpit and for the first time since the crash I felt slightly sick.

It was six o'clock the following evening, after further discussions when I finally got back to my home at Walton, driven there by George Edwards. The crash of Britain's first V bomber had been front page news to all the Sunday papers and Nancy, my wife, had had to cope as best she could with repeated questioning from many quarters. I could see that physically and mentally she was just about all in and I felt much the same myself. Fortunately our two children aged 6 and 2 were already in bed.

For several minutes while George Edwards talked to Nancy we still kept our guard up but when the front door finally closed I discovered I was not quite such a tough Glaswegian as I thought.

Jock Bryce was Vickers' chief test pilot between 1951 and 1964. He was involved in many first flights as pilot or co-pilot including the Viscount and the VC10 when he did the first take-off from the tiny Brooklands airfield and landed at Wisley.

Chapter One

FLIGHT DEVELOPMENT –
THE VALIANT AT VICKERS AND WOODFORD

The first Valiant prototype, WB210, with original intakes.
(*Aviation Historian*)

The brilliant account in the prologue of the loss of the first Valiant WB210 does not indicate what a real blow it must have been not only to Vickers, the manufacturers, but to the UK defence planners. The strategy of the UK government at that time was to counter the Russian threat with a nuclear deterrent delivered by high altitude aircraft capable of dropping an atom bomb behind the Iron Curtain. Incredibly by modern standards, three manufacturers were selected to produce machines capable of meeting the specification, Vickers, Avros and Handley Page; in fact the Vickers aircraft was chosen in order to get an aircraft quickly since the design was less advanced aerodynamically than the other two and therefore more likely to be delivered on time though the performance would not be as good.

The development of the three aircraft involved the country in a very large financial and technical risk and it must have required great determination not to cancel some or part of the programmes when WB210 crashed. However the second prototype was well advanced at the time of the accident and it flew on 11th April 1952 just under a year after WB210 and three months ahead of the original estimate in 1948, with revised production engine air intakes and having had the necessary modifications to prevent a similar accident. It is difficult to appreciate now the great pressure the Government was under at the time to produce a UK independent deterrent as soon as possible.

The Valiant design was very conventional in that very little sweepback was used on the aerodynamic surfaces which meant that the cruising speed at altitude was .78 Mach number compared with .86M of the other two V bombers; interestingly the maximum speed permitted in the pilots' notes at low altitude without underwing tanks was 360 knots compared with only 330 knots for the Vulcan Mk2. The forces to move the flying controls were very high and it was necessary to help the pilots by having four electrically operated hydraulic pumps; two for the ailerons and two for the elevators and rudders. Provision was made for continued powered operation if there was a partial electrical failure but, unlike the other V bombers, the aircraft could be flown manually without any hydraulic power in the event of complete electrical failure since the flight deck control hand wheels were connected directly to the control surfaces; however, as a personal reminiscence I remember testing a Valiant under complete manual control and the control forces were very heavy indeed.

Electrically the aircraft was similar to the Vulcan and Victor Mk1s in that there was not only a 28V DC electrical system but also a 112V system. The latter system was used for driving the control surface motors and also the undercarriage, flaps and airbrakes.

Like most new aircraft the Valiant had problems during development. One problem was with the cracking in the rear fuselage due to engine exhaust which took over a year to find a cure. Another problem was with the ailerons which were found to be prone to flutter approaching high mach number, .90M. Modifications were made on the first production aircraft to stop the flutter by introducing extra mass balancing on the ailerons but when testing at high mach number recommenced, aileron control was lost and the aircraft went into a spiral dive at about .85M with severe buffeting. The pilots managed to regain control using the rudder but both pitot heads were lost so that there were no airspeed or altitude readings. Brilliantly the pilots got the aircraft to Boscombe Down and managed to land safely. Stronger aileron rods had to be fitted to the whole fleet and eventually every production was flown to .92M though the service limit was .85M.

Speaking to Ted Dunne, a nav plotter who was in Dave Roberts's Grapple crew, he said when he was based at Wisley "max speed release to service of Valiant was 0.85M. However, whenever they descended from 45,000 feet to 38,0000 feet rapidly, I can't remember why, they always reached Mach 0.92!" He seemed surprised but that was what we did at Avros to clear the Flight Test Schedule with the Vulcan to make sure there was a 'safe' gap between the aircraft's extreme limit and the normal in-service maximum speed, although we hadn't had a history of aileron failures.

The first production bomber aircraft was delivered to the RAF in

Valiant photo-reconnaissance aircraft with cameras. (*Jerry Hughes)*

1954 two years before the Vulcan and three before the Victor. Later the aircraft was developed to fill several roles. Besides the basic role as a bomber, it was to be used as a photo-reconnaissance aircraft. Eleven photographic aircraft were produced which involved fitting a camera crate in the bomb bay, two extra cameras in the fuselage, another behind the camera crate and increasing the fuel capacity by extending the capacity of the two rear fuselage tanks. The first fully developed photographic aircraft was delivered in 1955 and the squadron was formed in 1956 at Gaydon. Later all the photo-reconnaissance aircraft were made flight refuelling capable.

The next role was as a tanker with the fitting of two large underwing tanks increasing the fuel capacity from approximately 56,000lb to 81,000lb depending on the fuel density. There was also provision for an auxiliary tank in the front of the bomb bay. The first unit to receive fully equipped flight-refuelling aircraft was 214 Squadron at Marham in 1959 under the command of then Wg Cdr, later Marshal of the Royal Air Force, Michael Beetham. 90 Squadron also became operational with tankers and the main use of these squadrons was to enable fighter

The Valiant as a tanker. The author is at the controls of the Vulcan.
(*A&AEE via Shaun Broaders*)

squadrons to be moved overseas without having to land en route.

Another role of the Valiant was radio countermeasures including high power jamming equipment. The first aircraft went to 199 Squadron at Honington in 1957 and then 18 Squadron at Finningley in 1958. There were a total of seven aircraft and the nav radar was replaced by another air electronics officer (AEO).

In order for the Valiant to be able to operate from a 2,000-yard runway at maximum take-off weight it was decided to fit rocket assisted take-off gear (RATOG), to the aircraft. The unit, called Super Sprite, produced 4,200lb thrust for up to 40 seconds and in Chapter Nine Milt Cottee describes a test from Boscombe Down in 1957 when the main spar broke doing a RATOG take-off. Although the units were cleared for use by the RAF and some squadron tests were carried out, there were a lot of failures and the project was abandoned in 1959.

Altogether 108 Valiants were built including one Mk2, WJ954, which flew on 4th September 1953, nicknamed the Black Bomber because of its paint scheme. The main change from the production Mk1 aircraft was that the retracted main undercarriage was in two fairings underneath the wing and the space vacated by the landing gear was filled with fuel tanks; the nose was lengthened by 4ft 6ins to keep the permitted centre of gravity virtually unchanged. However, there were no production Mk2s and the aircraft was able to be used to help the development of the Mk1 Valiant.

While the aircraft was being built and tested the RAF took the opportunity to send pilots and rear crew members to Vickers at Wisley to get experience of flying in the Valiant. This was very different from Avros at Woodford with the Vulcan where we just had one RAF liaison

Valiant Mk2 WJ954. (*Aviation Historian*)

pilot. Possibly one reason for the difference was that initially there was no flight simulator for the Valiant whereas the Vulcan pilots had a simulator from the outset.

Russ Rumbol was a nav plotter who flew with Vickers. He joined the Royal Air Force as a nav plotter in 1951 and was posted to 49 Squadron on Lincolns in 1953. He left in 1955 and joined 90 Valiant Squadron in 1956 winning the Lawrence Minot trophy in 1961. However he spent a few months at Vickers in 1956 before joining a Valiant flying in the Black Bomber and he has shared a few memories with us.

When I was with Vickers the Valiant had finished its development flying and was in the production phase. The aircraft had been flying years before I came to it in 1956 and was, indeed, in operational service at Suez, one flown by my future captain and friend, Philip Goodall. Nevertheless, although the aircraft type was not new, each machine to be flown had never before left the ground. Rather more than an air test, you'll agree.

We would take the aircraft from the hangar and fly off the runway at Weybridge, which was the old Brooklands race-track and very small indeed for a four-jet bomber being only 3,300ft long. We carried only the bare minimum to keep weight down. This meant four engines, two pilots, a navigator, a P-type compass and parachutes. Each take-off was accomplished with undisguised satisfaction and we had just enough fuel to get to Wisley, with its much longer runway of 6,600ft, which was only a few miles away. Even I could manage that bit of navigation.

Then followed a few weeks of testing every aspect of the aircraft performance before handing it to the users, the V Force squadrons. What did the navigator have to do? Well, not a lot. Certainly, we did not test the navigation equipment because none that was used on the squadrons, especially Navigation Bombing System (NBS) and Green Satin, was carried.

Really, he was there to keep the pilot out of mischief. No prohibited or danger areas, no unnecessary controlled airspace (remember, Wisley is pretty close to Heathrow) and no flying over the sea. At the end of the test, he gave a heading for base. Not difficult, eh?

How were we (two navigators at a time for a few months) selected? Was it a treat for being top of the course? By no means – we were pretty ordinary.

Frequently in the service, early courses produce a greater output than can be accommodated. In this case, more of us finished the NBS course than the OCU at Gaydon could handle. At the same time, Vickers Armstrong, although they had their own pilots and navigators, did not have enough to match the required production of Valiants. So, we were on loan for a few months and everyone was happy – rare in this imperfect world.

Such attachments are sometimes known as 'gash jobs'. This gash job was a dream posting. Financially, we were well off. As there was no service accommodation in the area, the Bomber Command postings staff kept us on what were called Rate Ones and Rate Twos, much more than we needed. We had digs in Weybridge, while Vickers treated us royally at lunchtimes, entertainments allowance being generous at that time. I think my favourites were Chateaubriand steak and Real Turtle soup, very politically incorrect.

We were usually able to do justice to the food for, if the test hadn't come up by lunchtime, the rest of the day was free. There was no night flying and it was autumn and then winter. It sometimes happened that there were very few flights during the week and then, miraculously and irritatingly, the aircraft was ready on the Saturday. Awfully suspicious minds we had, because we knew the ground crew were civilians and there was something called 'time and a half'.

As well as routine testing, we did other flights. There were trips in the Viscount and the 'Black Beast', the one B2 Valiant with performance exceeding the normal one. It came very close to Mach One, especially when Spud Murphy flew it. We also did some of the early in-flight refuelling trials, dry at the time. We were not always successful. After one flight when we failed to make contact, we were subject to sarcastic ribaldry in the crew room. A popular trip to the Paris Air Show was projected and the word went round "it's no use sending those two - they can't get it in!"

We were not directly involved with Suez but the fuel crisis hit us. Our digs were heated by oil and it was a mild winter. Also, we lost some of our free transport. Our problems were nothing to those of Vickers Armstrong. To run a test-flying establishment, you need a bit of fuel. They even had to apply for petrol for the petrol electric sets, which save aircraft batteries. From the civil servants, the reply came "use public transport".

Another highlight was a football match against Hawker aircrew at Dunsfold. The great Brian Trubshaw, who went on to be the chief test pilot for the Concorde, was our centre half. Though not a tall man, he was very powerful and a massive clearance from him hit me on the head and knocked me out. After that, my poor performance was attributed to this injury but the truth was that I wasn't very good.

As Magnus Magnusson could have said, "I'll finish as I started". For this match, Hawker provided the linesmen, one of whom was the fabled Neville Duke.

Vickers must have operated their testing in a different manner from Avros in that Russ tells me that very often he was the only crew member in the back whereas at Woodford we always had our own AEO to navigate with a Gee receiver and, more important, to control the electrics which really meant the 112V supply which powered the flying controls. Presumably Russ switched the generators on as each engine was started. I checked with Peter West[1] who was a Valiant AEO and then a Vulcan AEO; unlike the other V bombers there was very little the AEO could do with the electrical switches if things started to go wrong.

---

[1] See Chapter Fourteen.

The important thing if there was complete electrical failure was to shed loads like the flying control motors to conserve the 96V battery power and, certainly in the early days, this had to be done by the pilots. The Valiant's AEO main function was to operate the electronic warfare suite and HF STR18 radio if fitted; at Vickers therefore no AEO was required as long as the crew member in the back operated the generator switches and someone was responsible for telling the pilot where he was!

The other interesting point was that Russ inferred that it took some time to clear a production Valiant which certainly was very different from Avros. We occasionally cleared an aircraft in one flight but normally we needed one more and it took only a couple of days.

Dick Hayward was one of the early RAF pilots based at Wisley and unfortunately no longer with us. However we have some pages from his log book, one of which is attached in Appendix Two. In addition Keith Walker was his co-pilot and fortunately he has written some stories about Dick in Chapter Seventeen. Clearly Dick, unlike Russ, was flying on all sorts of development tests, preparing the Valiant for all its various roles. Looking at the 'tail numbers' most of the aircraft were in the CA fleet, that is aircraft that were allocated to the aircraft manufacturers for testing by the Controller Aircraft. So in 1956 Dick flew in many different test aircraft that were

Dick Hayward in the lead Valiant on 13th June 1957, flown to celebrate the Queen's birthday.
(*Lesley Hayward*)

developing flight refuelling, aerodynamics vortex generators, de-icing trials, performance measurements, generator loads, RATOG Sprite jettisoning and autopilot instrument landing system (ILS) trials.

The autopilot ILS flights in particular caught my eye as the aircraft WP208 must have been delivered to Boscombe mid-July. I was a test pilot at Boscombe and I flew it on autopilot acceptance testing at the end of July. In fact it was my last flight in the Royal Air Force before joining Avros.[2]

Examining Dick's log book a little more, my memories came flooding back because being a test pilot at Boscombe Down at the time flying the Valiant I recognised all six Vickers pilots with whom Dick had flown; Bill Aston, Dizzie Addicott, Jasper Jarvis, Spud Murphy, Staff Harris and Brian Trubshaw. I knew them all and had flown with several of them. We all used to meet at the Farnborough air display every year but Spud I knew particularly well as he joined Handley Page and we were in competition, the Avro 748 against the Handley Page Herald. Brian Trubshaw offered me a job testing the Concorde which was very tempting but I declined; however he later recommended me to become technical board member on the Civil Aviation Authority which was much appreciated.

One forgotten and important role for the Valiant was to help develop Blue Steel, an air-launched peroxide-powered missile designed and developed by Avros on the other side of the airfield from where Avros did the test flying; the Valiant carried the weapon to enable testing of all the electronics on board though only the Victor and Vulcan carried the weapon in anger. In fact it was one of my first jobs when I joined Avros as a test pilot to fly and train the pilots who were to fly the test Valiants. The Valiants then went out to Woomera in Australia to carry out more Blue Steel testing. **John Saxon** was a test engineer for Elliott Automation that provided the inertial navigator system for Blue Steel and he describes the work he and the Valiants did. John's initial job at Woodford was clearly to analyse the Woodford Blue Steel Valiant flight testing.

All the initial test flying of the air-launched weapon was done using Valiants WZ370, WZ375 and WP206. My background is in electronics engineering and in early 1960 I found myself unemployed as the American firm I was working for in the UK went belly up. So I went to the Government employment office who sent me to Elliott Brothers in Borehamwood for an interview. They asked me what I knew about inertial navigation systems, "absolutely nothing" I replied, and the response was, "no problem, no-one else knows anything about them either". They then gave me a job which led to joining the Blue Steel missile flight trials team due to go to Australia. Nothing can beat low unemployment.

We worked initially at Avros up at Woodford – home of the production lines for Vulcans. Very impressive and very noisy. My main job was to be in the maths analysis group to crunch the numbers to decide the success or otherwise of the launches above the Woomera range in Australia. So we learnt our trade using a mix of dummy and actual lab data.

---

[2]　www.blackmanbooks.co.uk/valiantbydate.htm

Finally the great day came and we decamped and travelled to Adelaide on one of the RAF regular flights. Quite a contrast between Salisbury Plain with lots of snow and the first stop in the middle of the Sahara desert! It was a very nice way to travel, taking a week to do the journey.

Once established in Adelaide it was apparent that some lucky people from the team would get to fly on launch and other trials. As my group had to analyse the resulting data, who better to ensure that everything was conducted to collect the data with the greatest accuracy and precision? I made my pitch to join the flying team, and somewhat to my surprise I was accepted.

I don't know how nice the Valiant was to fly from a pilot's perspective as my flying was limited to the centre rear 'nav plotter' position but I had no complaints. The fact that I was able to fly in the Valiant at all (and the other two V bomber types) as a civilian was totally unexpected.

We started with three Valiants – the real work horses of the Australian trials. They were used for both 'carry over' tests to prove new hardware configurations and possible launch modes, as well as the early launches of full size missiles. We received a Vulcan and a Victor B2 later, but from our records there were a total of 274 flights in the trials and 147 of them were Valiant flights, 69 Vulcans, and 58 Victors. At least thirteen launches were made from Valiants in Australia[3].

The picture below shows one of our Valiants taxiing towards the Edinburgh Field 'loading bay' which was surrounded by a massive chain link fence. The

Valiant carrying Blue Steel at Edinburgh Field. (*John Saxon*)

---

[3]   http://www.blackmanbooks.co.uk/blue%20steel%202.htm

fence would probably stop some pieces from destroying nearby buildings if things went bad during the missile loading process. Luckily we never did knowingly fly with active atomic warheads. If you look carefully at the picture you can just see that the aircraft is carrying a full size Blue Steel.

I enjoyed the operational aspects of the flights in all three V bomber types. Setting up the next fix points, gradually compensating for gyro drift errors, keeping detailed written logs in sometimes difficult conditions, taking photos of the instruments at specific times, starting recorders etc. We often compared launching a Blue Steel to launching an ICBM or large rocket, but from a moving launch platform travelling five miles up.

The launch countdown was a real team effort, everything had to be completed perfectly by each crew member, and no countdown holds were allowed. The launch point had to be met accurately in position and time approaching at 600+ mph. Most of the airborne equipment we used was analogue.

I returned to the UK to learn about high altitude flying in the Victors and Vulcans since in the Valiant we never went above 45,000ft; in fact half way through the trials the UK nuclear deterrent went low level and so did our Blue Steel launches.

In the Farnborough decompression chamber we learnt how to 'pressure breathe' where oxygen is forced into your lungs at altitudes above 46,000ft. It is the reverse of normal breathing and quite hard work. The highlight of the chamber training was an 'explosive decompression' from 8,000ft to 56,000ft in a couple of seconds. The air goes white with condensation and life jackets and other things inflate as any remaining gases expand to ten times or so their normal volume. But they didn't warn us that all our internal bodily gases also expand by the same ratio. Didn't seem too bad in the chamber, but one certainly realised just how bad it was when we eventually got a sniff of the outside air!

I spent nearly six months in the UK during that trip. During that time I flew some sorties from Woodford, mostly at night as we were testing star sights in preparation for the possible equipping of the V Force with the Skybolt missile system. It's quite possible that Tony and I flew some of those sorties together – but no records exist of who occupied the rear seats. I do remember that on return to Woodford one night it was socked in, but Manchester Ringway was OK. We were not carrying an INS which might possibly have helped in navigating the Ringway taxiway.

The return to Australia and back to the trials was great! I had been missing the weather and the laid-back life style. But after a bunch of Vulcan and Victor flights all good things have to eventually end, and I was returned to the UK to work on the TSR2 project.

Chapter Two

OPERATION TOO RIGHT

Valiants en route to New Zealand. (*David Sykes*)

David Sykes was a Halton Apprentice and served in the RAF from 1954 to 1966. He primarily worked on Valiants at 232 OCU Gaydon and then went on to the Vulcan. On leaving the service he obtained an HNC in electronics. By chance our paths crossed since he joined Smiths Industries after leaving the RAF where he worked on the Smiths Military Flight System, the Vulcan auto throttle and the auto-landing autopilot when I was using the equipment and clearing the whole installation on the Vulcan Mk2 at the same time.

Operation Too Right was the code name for the first overseas flight carried out by RAF V bombers and which featured two Valiant B Mk1 aircraft WP206 and WP207 of 138 Squadron based at Wittering, UK. Valiant aircraft were the first of the V bombers to enter service and at the time of Operation Too Right it was the only operational V bomber because the Victor and Vulcan had not yet been delivered to the RAF. The two Valiant aircraft taking part in Too Right were under the immediate command of Squadron Leader R G Wilson DFC, the captain of WP206 and the operation was under the overall command of Air Officer Commanding-in-Chief, Bomber Command, Air Marshal Sir George Mills who accompanied the tour with his wife, Lady Mills, flying in their distinctive VIP Hastings. The final destination of Too Right was Christchurch, New Zealand and all aircraft, consisting of the two Valiants, the VIP Hastings and four other Hastings transport aircraft which carried the ground support crews, had a planned arrival at Harewood Airport, Christchurch on Monday 19th September 1955.

Each Valiant had a crew of five with a sixth seat fitted to accommodate the crew chief. The five aircrew consisted of first pilot, co-pilot, radar navigator, radar plotter and AEO. The crew were accommodated in a pressurised cabin at the front of the aircraft, with the pilots being seated in ejector seats and the rear crew members relying on manual evacuation through the access door. The crew chiefs were in charge of the replenishment and technical servicing of their particular aircraft and both held the rank of chief technician.

Although WP206 and WP207 were 138 Squadron aircraft, they had been prepared for the trip at 232 OCU RAF Gaydon and the ground crew selected for Too Right were also domiciled at Gaydon and were not drawn from 138 Squadron personnel. The reason that Gaydon tradesmen were chosen was possibly due to the fact that there were no operational squadrons at the time; the initial planning for Too Right took place at Gaydon and so it seemed logical to use the experienced Gaydon personnel who were available. Gaydon had been operating the Valiant for some months whilst aircrews were being progressively trained to enable the first operational Valiant squadrons to be formed and it so happened that Operation Too Right took place during the time when 138 Squadron was building up to become the first operational squadron. In recognition that Too Right was essentially a 138 Squadron operation the ground crew travelled from Gaydon to Wittering by bus on Friday 2nd September and were billeted there for one or two nights. They progressively flew out from Wittering in the four Hastings aircraft on a staggered schedule during the night of Saturday 3rd and early Sunday morning of 4th September 1955. After the completion of the tour several ground crew were posted out to 138 Squadron and it could be argued that the remainder, who stayed on at Gaydon, could be considered as being honorary 138 Squadron members after spending several weeks servicing these 138 Squadron Valiants.

Ground crews were positioned along the route by the four Hastings aircraft from 40 Squadron, Transport Command, which were manned by mixed RNZAF and RAF flight crews. These four aircraft carried a full range of Valiant spares, tools and servicing plant between them and they also carried their own ground-support personnel. Each trade group was headed by a corporal and work on each individual aircraft was controlled and overseen by the respective crew chief. The crew chiefs were Chief Technician R V (Pat) Patrick of WP206 and G (Johnnie) Greyburn of WP207. We were also accompanied by Master Technician D G (Doug) Livett, who was in overall charge of Valiant servicing, documentation and administration. There was also an equipment and supply officer on one of the Hastings but he was adversely affected by the climate and was evacuated back to the UK in the early stages of the trip. He was not replaced.

As a corporal at that time, I was fortunate enough to have been chosen to oversee the instrument trade which consisted of two junior technicians as well as myself. It was planned that the Hastings aircraft, complete with servicing crews etc, would be staggered along the route in order to keep pace with the faster

Valiants and my part in the plan was to be on the last aircraft in order to be in position at RAF Habbaniyah in Iraq, for the first stop of WP206 and WP207. The other three Hastings pressed on ahead and one was in position at Maripur near Karachi; the next at Negombo in Sri Lanka and the last at Changi in Singapore where we would all gather to carry out primary servicing on the Valiants and where we would take a short break for shopping, sight-seeing and relaxation. After staying in Singapore my particular Hastings flew on to Darwin in Australia, then on to RAAF Amberley near Brisbane, from there to Melbourne followed by Edinburgh Field near Adelaide. Other Hastings, with their servicing crews, were positioned in Sydney and Perth and one or other of the Valiants were put on static display at the main airports. They also did flying displays and flew over each city at a pre-arranged time so that excited school children and other groups were able to assemble and enjoy the show. All aircraft then flew from Edinburgh Field to Harewood Airport, Christchurch, with air and ground crews staying at RNZAF Base Wigram. Following on from this, we moved on to Ohakea near Wellington and Whenuapai in Auckland, where programmes similar to those in Australia were completed.

At Habbaniyah we were waiting for the Valiants to arrive when suddenly we saw WP206 streaking overhead, at high altitude, in the process of breaking the London to Baghdad record. After both aircraft had landed and parked, we proceeded to place bungs in the various intakes and replenished and serviced them ready for the next stage of the trip. Whilst this was happening, a violent sandstorm blew across this desert airfield but, fortunately, it was just as we had completed our tasks and when the Valiants left for Karachi at about 1 am in the morning, followed closely by we tired individuals in our Hastings, all was remarkably calm.

After we had been sitting sleeplessly for some time on our noisy aircraft, our Hastings captain received a signal that both Valiants had aborted. He reported that WP207 was back at Habbaniyah with pressurisation failure and WP206 had made an emergency landing at RAF Sharjah, in Trucial Oman, with a disintegrated engine. As a result we diverted to Sharjah and had breakfast after off-loading the engine fitters and a spare engine; we then flew back to Habbaniyah to sort out WP207.

When we arrived at Habbaniyah we found that the 'snag' had already been fixed by the crew chief, the fault being due to the ingress of sand in the combined valve unit, most likely caused by the sandstorm. After seeing WP207 off, again at about 1 am, we once more piled on board our Hastings, en route for Sharjah and with little prospect of sleeping due to the excessively noisy aircraft. We arrived about dawn after two days and nights of almost no sleep and were met at the opening door by the crew chief of WP206 who barked, "Everyone report to WP206 immediately!" As we stormed past him like brainless zombies, one member of our party had the temerity to reply "Fuck off!" but had the presence of mind to add "Chief!" respectfully at the end of his insubordinate outburst.

Fortunately, it did the trick and we left a flabbergasted crew chief with mouth wide open as we staggered past and onward to the nearby transit block, where we crashed onto the nearest beds we could find. After several hours of badly needed sleep, in a beautifully air-conditioned room, we started work on WP206.

Engine change, Sharjah. (*David Sykes*)

For the next few days we were involved in changing number three engine, working through the middle of the hot, steamy day when everyone else had their siesta, playing sport or going to the nearby beach. The Valiant was a superb aircraft on which to change an engine as no crane was needed. The engine was lowered on winches, which were mounted on a flat beam on the main plane. We had the spares and winches but not the beam which, we found, was on one of the other aircraft on the way to Singapore. It was decided that it would be quicker for a Canberra to fly a beam out from the UK in its bomb bay and some hours later the Canberra arrived complete with the desperately needed beam. After successfully changing the engine, the aircraft was painfully and repeatedly refuelled by a group of excitable natives from a three-wheeled bowser, which was similar in size to a ride-on mower. It was then realised that we had no 112V starter set and so, to solve this problem, every available battery was snatched from the small fleet of Mechanical Transport (MT) vehicles and any other sources until we had 112 volts to start the engine! We managed, with fingers crossed and after one agonising failure, to get the Valiant started on one engine, from which the other three engines were serially started. After completing engine test runs, WP206 took off from the Sharjah runway, which was appreciably shorter than that recommended in the operation manual.

Whilst working on the engine the crew chief told me all about the in-flight emergency and said that, suddenly, there was an enormous bang followed by a fire warning in number three engine. The captain gave the order to abandon aircraft and the crew chief strapped on his parachute, pulled his emergency oxygen supply and was just about to blow the door when, over the intercom, the captain then cancelled the order. He declared that the aircraft was handling OK and that he had extinguished the engine fire. The crew chief said they were all

Engine change completed at Sharjah. (*David Sykes*)

very relieved because the thought of splashing down in the shark-infested Indian Ocean, over which they were flying, was not a happy one.

After rectifying the two mishaps, both aircraft performed perfectly for the rest of the trip. We were feted wherever we went and the Australian and New Zealand newspapers covered the event with multiple articles and photographs but, strangely, it is almost as if this trip never took place as far as historic records in the UK are concerned. In strong contrast, much has been written about the trip by the Vulcans the following year, which culminated in a tragic accident at London Airport. There was also a mishap with a later visiting Vulcan

Night stop, Idris. (*David Sykes*)

which made a heavy landing at Wellington Airport and subsequently made an emergency landing at Ohakea, which was also well recorded. In the book entitled *V Bombers*, by Robert Jackson, the well-known aviation author, it is written: 'In June 1956, Valiants went overseas for the first time when two aircraft flew to Idris, in Libya, to take part in Exercise Thunderhead'. I wrote to the publisher advising of the error and received a brief and polite reply, but I could be excused for feeling that I detected a subtle 'OK, Smart Arse!' sort of tone about the letter! We did have a serious case of misconduct during *Too Right* by one of the aircrew which compromised air safety. Could it be that this incident was effectively covered up by ensuring that all publicity was kept under wraps and the press kept away to prevent the story from emerging?

The Valiant was the first V bomber in service and had some unique features and an excellent record but, sadly, the type met a very sudden demise due to structural failure. This aircraft was used as a conventional bomber in the brief Suez campaign and was the only V bomber to drop nuclear weapons in the trials at Maralinga and Christmas Island. I believe that the Valiants were broken up with indecent haste with the result that only one example (XD818) now survives at the RAF Museum in Cosford.

David Sykes i/c instrument servicing, Junior Tech; Cyril Sheppard
(engine fitter), Junior Tech; Terry Tranter (air wireless fitter) and
LAC Pat Yearsley (air wireless mechanic) in New Zealand. (*David Sykes*)

Chapter Three

OPERATION BUFFALO –
ATOMIC TESTING IN AUSTRALIA

The decision was taken by the UK government in 1946 to develop an atom bomb as a nuclear deterrent and have it delivered by a 'long-range' bomber aircraft behind the Iron Curtain. This meant that in effect in the late 1940s and in the 1950s there were two parallel development projects taking place, the atomic bomb and the aircraft programme. As a result, the scientists were designing and manufacturing the bombs at the same time as the V bombers were being built but only the Valiant, the first of the V bombers, actually dropped live nuclear weapons.

This chapter tells the story of the weapon development and the Valiant bomb drop in Australia but in writing the account it is difficult not to be struck by the way social attitudes have changed through the years. At the time the urgency to carry out the tests seemed to have taken priority over the people who lived at the development sites; in fact, it would seem that even the workers constructing and operating the test locations may not have been adequately protected from the consequences of radiation and Appendix One of this book discusses some of these issues.

The nuclear scientists needed locations to test their development triggers and weapons and finding suitable locations was not easy. The Americans were not co-operative in providing a site but the Australians gave permission to use the Montebello Islands and Emu Field in the Great Victoria Desert of South Australia. So on 3rd October 1952 the UK carried out its first atomic explosion; the explosive device was put inside a naval frigate, HMS *Plym*, which was completely destroyed. A year later the first nuclear tests on the Australian mainland were carried out at Emu Field on 15th and 27th October where test explosions were from a 31m steel tower. However neither sites were really suitable for proper measurements and so on 30th October 1953 the British formally requested a permanent test facility from the Australian government; under the then Australian Prime Minister Sir Robert Menzies, it was agreed that the recently surveyed Maralinga site in the Woomera Prohibited Area of the remote western area of South Australia could be used for testing as a joint, co-funded facility between the two governments. Each test would need to be agreed and no fusion/thermonuclear weapons could be used over Australian soil. Consequently for the later tests with the high yield thermonuclear explosions, called Operation Grapple, the UK used and developed a remote island site in the Pacific, Christmas Island with the drops at Malden Island 400 miles to the south, both islands in the former British colony of the Line Islands and now part of the independent Republic of Kiribati.

The tests at Maralinga were called Operation Buffalo and consisted of four tests, the fourth of which was to be an air drop from a Valiant; the other tests were from a 31m tower and a surface explosion.

The Valiant, the first of the UK's three V bombers, was only just entering service at the time the UK was ready to test its own atomic and nuclear bombs. The newly re-formed 49 Squadron was the unit chosen to participate in these tests. So in February 1956 C Flight of 138 Squadron moved to RAF Wittering and finally became 49 Squadron on 1st May of that year. The embryonic squadron, initially three Valiant B1 aircraft and crews, was commanded by Sqn Ldr Roberts and then from September by Wg Cdr Hubbard.

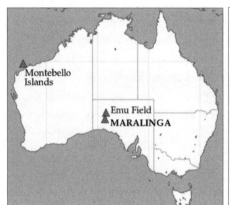

The test sites in Australia and the Pacific Ocean.

The squadron was tasked with carrying out trials leading first to a live drop of 'Blue Danube', Operation Buffalo, and then going on to perform trials of Britain's thermonuclear megaton weapon, Operation Grapple. The two 49 Squadron Valiants designated for Operation Buffalo, WZ366 and WZ367, left Wittering early in August 1956 for the Maralinga Range in Australia.

The captain of WZ366 was Sqn Ldr Ted Flavell and the co-pilot was Flt Lt John Ledger and we have access to both their logbooks[4]. Ted was a Boy Entrant 37th Entry to RAF Halton, passing out as an aircraftman fitter 1st class, and went to RAF Northolt. He applied for aircrew training and was sent to Canada. On return to England there was a variety of roles. First flying Whitleys for live para drops at No.1 Para School at Ringway; then flying Whitley, Stirling, Albemarle and Halifax aircraft during World War II; finally Albemarles for SOE work and for glider towing including D-Day and 2 lifts to Arnhem (Operation

---

[4]   See Appendix Three.

Market Garden). Halifax work included SOE to Norway and also some bombing missions. After the war Ted was posted to Palestine again on the Halifax. Back in England he then went to the Airborne Forces Experimental Establishment at Beaulieu where many types of aircraft were flown. After a short posting to a glider station at Dumfries, Ted was returned to active duties on Wellingtons and then Lincolns at Binbrook. It was here that he did a conversion onto the English Electric Canberra followed by conversion to the Vickers Valiant at Gaydon and Wittering. This period of flying saw Ted's selection for Britain's atomic programme.

Ted and John appear to have been together on 138 Squadron since 16th November 1955 when John joined the Valiant force from 232 OCU. John completed five hours flying in two trips on No 1 Second Pilots course, 232 OCU before joining 138 Squadron. On 49 Squadron, the crew flew WZ366 from May 1956 on a series of trial flights and dummy runs; trial F4Ds are recorded in his logbook on five occasions during May and June 1956. These were followed by several visual bombing sorties dropping 100lb bombs at Sandbanks, Jurby, Chesil Bank and Wainfleet.

The transit route out to Maralinga in Valiant B1 WZ366 was as follows with the return two months later:

Date	Route	Time
5 Aug 56	Wittering to El Adem	4.00
6 Aug 56	El Adem to Mauripur	5.50
8 Aug 56	Mauripur to Negombo	3.25
10 Aug 56	Negombo to Changi	3.55
13 Aug 56	Changi to Darwin	4.30
14 Aug 56	Darwin to Edinburgh	3.20

It is interesting to note that Gan in the Maldive Islands was not available at that time so that the routing was via Mauripur. RAF Mauripur was a Royal Air Force airfield near Karachi which was still used by the RAF after the creation of Pakistan in 1947. Negombo was in western Sri Lanka, or Ceylon at that time. On a personal note when I ferried a Vulcan out to Edinburgh a few years later I was able to use El Adem-Aden-Gan-Changi which would have been the preferred Valiant route.

Before the live drop the crew flew a series of practice and telemetry calibration trips from Edinburgh, landing at both Woomera and Maralinga, a total of twelve trips. These started on 23rd August 1956, culminating with the first atomic bomb drop. John says they dropped a series of dummy weapons prior to the live drop. These were probably on 29th September 1956 – 'HE3 as briefed' in his log book and 4th October 1956 – 'Round 2 as briefed'. Round 3 was the live drop, a sortie that lasted 1hour and 35 minutes.

Interestingly, John Ledger has no recollection of any particular debate as to whether a ground site in Australia should be used for

Blue Danube bomb.

atomic explosions. The crew did think that it was strange to be permitted to drop such a powerful weapon over land with the potential risk of fallout being spread over an inhabited area. This was particularly the case later when the wind changed direction and caused the fallout to be blown over the southern towns.

The weapon being dropped was called Blue Danube and was similar in size and shape to the Tallboy HE weapon of WWII, but had an operational explosive yield of 40 kilotons. The original plan was to use a standard bomb, fused to detonate at 1,200 feet. However, if the fusing system failed, a surface burst of 40 kilotons was unacceptable so the bomb was modified to give a three-kiloton yield with a burst altitude of 500 feet. The first British nuclear weapon was dropped from WZ366 on 11th October 1956. Captained by Sqn Ldr Flavell, the aircraft let go the weapon from 35,000 feet, visually aimed after a radar controlled run-up. Telemetry confirmed a burst height between 500/600 feet, 100 yards to port and 60 yards short of the target. Both Sqn Ldr Flavell and his bomb aimer, Flt Lt Stacey, were awarded the AFC.

John Ledger stated that they wore normal flying suits with no additional protection. He remembers that there was very little advice offered pre-flight although post landing, after the bomb drop, there was a team to meet the aircraft to assess damage and radio-active contamination. John remembers the scientists got a little excited about the crew door surrounds, but that was all.

With regards to the special modification state of the aircraft, WZ366 had been fitted with windscreen blinds (John thinks made of steel) that obliterated all view from the cockpit. There was a small central slot in the blind that allowed the crew to see out for take-off and landing, which was then closed after take-off. This would explain the emphasis on instrument flying during the build-up and training phase in the UK.

John thinks that the A bomb was dropped from the visual bomb aimer's position with the navigator then scrambling back to his seat post release. The crew seemed to know about the initial shock wave which he describes as moderate to severe turbulence. However, they were less prepared for the severity of the return wave, which caused them some consternation. John said that the navigator did not get back into his seat in time and was thrown about the rear cockpit due to the blast wave, much to the amusement of the rest of the crew.

Contrary to other accounts, John has always been adamant that only the normal crew were on the aircraft during the actual bomb drop with no additional observers or senior officers.

John went on to gain a captaincy on the Valiant in late 1958 and served on 49, 138 Squadrons and 232 OCU until December 1964 when he converted to the Britannia.

Bill Evans was a corporal airframe technician supporting the Valiants. He was involved with the Valiant from fairly early days when he was sent to the Vickers aircraft factory at Weybridge on the first ground engineering course as an airframe fitter (not crew chief) in 1954. On completion of the course he was posted to RAF Gaydon to carry out acceptance checks and schedule servicing on the Valiants prior to them being allocated to squadrons.

I volunteered to take part in the nuclear trials in Australia and then the thermonuclear trials (Grapple) at Christmas Island and was subsequently posted to RAF Wittering 1321 Flight which later became 49 Squadron. I worked with the Valiant doing Operation Buffalo in Australia and then on to three of the four Grapple tests at Christmas Island. I worked with the Valiant until 1959 when I was posted to Singapore for two and a half years. On my return to the UK I was posted back to Valiants at RAF Wyton (543 Squadron) and worked with them until their demise when I went over to the Victor.

For Operation Buffalo the 49 Squadron ground crew left RAF Wittering on Saturday 4th August 1956 in a Transport Command Hastings for Australia via Idris, North Africa, El Adem (Libya), Habbaniyah (Iraq), Karachi (Pakistan), Negombo (Ceylon), Changi (Singapore), Labuan (Borneo), Darwin, Edinburgh Field (Adelaide), and finally Maralinga. We were supporting two silver-painted Vickers Armstrong Valiants: WZ366 captain Sqn Ldr Ted Flavell and also the CO of the detachment and crew, and WZ367 captain Flt Lt Bob Bates and crew.

On Monday 6th August, third day out of the UK we landed at Habbaniyah. As we walked down the steps at the rear of the Hastings the heat hit us and at first we thought it was coming from the engines until we realised that it was actually the outside temperature. This was a huge difference to our bodies having come from inclement weather only three days previously. Luckily we were only there for twenty-four hours, most of which time we spent under a cool shower. A nasty experience was encountered when we met up with a camel spider in the toilet block, it was as large as a man's hand, and we left very rapidly. The next day we arrived at Karachi at the civilian airport, we were then bussed to the Pakistan Air Force base of Mauripur situated on the outskirts of the city where our two Valiants had landed. We passed through a large area of slums and horrible vile smelling parts of the city; there had been torrential rains and downpours a week to ten days prior to our arrival and floodwater was everywhere – a stinking bluish grey colour with the inhabitants wading through it or sitting forlornly on tables and boxes above it in tumbledown shacks – a very sobering experience. The billets at the base were low concrete huts with small open slits for windows and air, very medieval. We were all housed in one large room with beds only – no chairs, tables or anything else – but there were three electric ceiling fans to churn the humid air. Darkness fell early and with little to do and nowhere to go out to we just had to make our own entertainment, part of which was an all-in pillow fight which was abruptly stopped when a thrown pillow knocked a blade off one of the fans causing it to rotate in a very unstable manner – everyone ducked and weaved to avoid it; result we were down to two fans, not good!

During the evening we were visited by a local old chap carrying two large wooden boxes on a long pole over his shoulder; one contained a small gas stove and horrendously strong hot coffee, the other held some rather dubious-looking brightly coloured cakes; we settled for the coffee. Next morning the Valiants were away after pre-flight checks; the runway was covered in residual rainwater and they got airborne through massive sprays of steam and water rather spectacularly but safely.

At Negombo we had a hiccup in our journey when WZ367 had an engine failure on start-up (seized engine). Departure was delayed for seven days until the arrival of a new engine from the UK so we enjoyed swimming in the station pool and visiting the town. Interestingly, the cause of the engine seizure was an old type twelve-sided three-penny piece which had escaped from the corporal engine fitter's trouser pocket unbeknown to him when he carried out the pre-flight checks to the engine intakes. This coin is now on display in the Rolls-Royce aircraft museum.

On Thursday 16th August we arrived at Changi which was very busy and made us pine for our previous relaxed week! We spent two days in Singapore during which time we visited the city – quite an eye opener. A couple of us went to the 'Happy Dance' dance hall, again quite an experience. It's the first and only time in my life I have paid to dance with a female, one dollar for three dances. On the Saturday we departed for Darwin stopping at Labuan to refuel and lunch. We arrived at Darwin just after midnight, a meal was provided – leftovers from their supper. The accommodation at Darwin Air Base consisted of wooden huts built on concrete pillars approximately ten feet above ground. We were billeted in makeshift accommodation under the huts with hessian sacking wrapped around the pillars and very basic camp beds to sleep on. Fortunately this was only a short stay and then a flight from Darwin to Edinburgh Field near Adelaide, from top to bottom of Australia following an arrow straight track out of Darwin passing Alice Springs about halfway down, and straight on again, nothing to see except outback bush and huge dried-up salt lakes, the Great Australian F**k All.

We finally arrived on 20th August to find what a joy Edinburgh Field was, wonderful accommodation, good food and back with the living once again. We even had electric washing machines in the billets. However, the weather was cold and wet. We spent three days and nights here and then flew with the old Hastings up to Maralinga on the Thursday. We were accommodated in tents at the side of airfield near our aircraft dispersals. Lighting in the tents was from hurricane lamps only. We were approximately three miles from the main camp and had a rather makeshift corrugated mess hut which was shared by officers and other ranks; the food was superb and plenty of it. On the Monday we had a social evening with our aircrew and officers and played liar dice and darts; it was a really good evening drinking mainly sweet and dry sherry because the bottled beer was prohibitively expensive. We all felt a bit under the weather the next day. However, as there was very little flying we spent the time over the next three days improving our tented accommodation. We also replaced two cracked cabin windows in WZ367.

On Saturday we returned to Edinburgh Field and stayed there for a week doing very little except socialising. Then back to Maralinga again to settle in once more. The first atomic bomb test, a ground one in the series, was due Tuesday 11th September but it was postponed owing to adverse wind direction; in fact this test was delayed for another fifteen days because of bad weather conditions; at one time the test came within two minutes of firing, a most frustrating time.

We returned to Edinburgh Field and stayed there for a week doing very little again. The first ground atomic test finally took place on Saturday 27th September and was carried out at 17.00 hrs. We were standing out in the open on the airfield beside the runway with our backs to Ground Zero which was approximately fifteen miles away. We were given the countdown and were informed we could turn and observe the results. There was a mushroom cloud rising rapidly into the air surrounding a fireball, it was quite a time before we heard the explosion and there was very little shock wave apparent.

Two days later both Valiant aircraft dropped practice high explosive bombs with excellent results. When carrying out the after-flight inspection on WZ367 I discovered a six-inch split in the underside skin of the starboard inner flap, cause unknown. As we had no repair material with us I cut a piece of metal from the crew room hut and carried out a 'temporary' cover patch repair with 'pop rivets', the intention was to do a more substantial repair when returning home, however, this never happened and years later the crew chief, Denis Whitfield, told me the aircraft had gone to its grave with the patch still intact!

On Sunday 30th September we went up to the forward test area by Land Rover to see the results of the airbursts and also to within approximately half a mile of the atomic explosion; we were instructed not to get out of the vehicle but observe only because we were only wearing khaki drill shorts and shirts. On the following Thursday the scientists carried out the second atomic test at 4.30 pm which was not as impressive as the first – no fire ball, just the usual mushroom cloud.

On Saturday 6th October we went to an open air cinema at the main camp and saw the film *Jumping Jack* with Dean Martin and Jerry Lewis, Sir William Penney was there, first time I had seen him 'in the flesh'. Then the following day I assisted boffins in the forward area to set up and clear tracks for rockets.

Finally on Thursday 11th October RAF history was made with the first live atomic bomb drop from a British Vickers Valiant aircraft — WZ366 captained by Sqn Ldr Ted Flavell.

Britain's first air-dropped atom bomb.

On Friday 12th October we went back to Edinburgh Field again. There I learned I had been under paid £1 0s 1d per week for quite some time. Feeling rich, I bought an engagement ring in Adelaide. We spent the next few days packing for our return home. On Thursday 18th October we departed for Darwin by Hastings aircraft for an overnight stay. Next day we took off for Changi over the Timor Sea but had to turn back after twenty minutes due to one engine overheating (a cooling door operating lever had sheared); we ground crew on board were rather on edge at this stage realising something was amiss because the engine was revving up and down and the aircraft was doing a tight turn back to Darwin. At the same time what looked like smoke was streaming back from under the wing; this turned out to be fuel being jettisoned. We were all rather acutely aware that the Timor Sea had more sharks per square yard than any other sea. However, all was well, we landed safely and with the help of our engine fitters we were soon repaired and on our way again refuelling at Borneo en route. On Saturday 20th October my diary tells me: 'changed money and went shopping in Singapore, returned broke. While there saw a dead Malayan in the street'.

Next day off to Negombo calling at Butterworth airbase in Malaya to refuel the Hastings. WZ367 had a problem at Negombo but it was soon fixed. Monday 22nd October we were off to Mauripur where we stayed overnight. The Hastings was able to land at the base as conditions were much drier than on the outward journey. On Tuesday we went to Habbaniyah but it didn't feel anything like the heat on our outward journey. Went for a swim in the station pool, temperature of the water was 70 degrees. We thought it rather cold. Wednesday off to Idris but were diverted to Luqa, Malta due to strikes at Idris. Finally on 25th October we reached the last stage for return to the UK which proved to be a rough and bumpy trip. Arrived Wittering safe and sound, had to pay £7 10s 0d to customs for the aforementioned ring which was for my fiancée, Rosemary, and future wife. She still has the ring after 55 years. Then off home for a much awaited leave of one week.

I feel I must make a comment on Bill Evans's account of ground supporting the Valiants. I've had lots of splendid stories by aircrew and ground crew on the other V bombers and the Nimrod but this is the first one that is clearly related from a diary which gives it such an air of authenticity.

While the testing was going international pressure was growing to stop weapon testing and this put pressure on the atomic scientists to develop the new weapons as quickly as possible. The aim was to have a fusion thermonuclear bomb and it was decided to do some more tests at the Montebello Islands called Mosaic in parallel with the Buffalo tests. The first one from a 30m tower was on 16th May 1956 which had a yield of 15 kilotons with very little fusion. Consequently a second test was carried out on 19th June 1956 but the yield was much greater than expected, 98 kilotons. This was the highest yield test yet and apparently broke an assurance made personally by PM Anthony Eden of the UK to PM Robert Menzies of Australia that the yield would not exceed 62kt, which was 2.5 times that of the first test at the Montebello Islands and the true yield was concealed until 1984.

To complete the Australian story, the following year in September/ October Maralinga was used for Operation Antler where three explosions were carried out developing the thermonuclear fuse.

Chapter Four

OPERATION MUSKETEER – SUEZ 1956

This was a confrontation that was generally referred to, among other names, as the 'Suez Crisis'. It stemmed from General Nasser's decision to nationalise the Suez Canal which was not in the interests of the West, who still wanted control of the canal, and thereby wished to remove Nasser from power. Furthermore, Nasser's decision also brought home the problem that the West would face with regards to access to oil from the Middle East. The contenders were Britain, Israel and France on the one side and Egypt on the other. The USA, Soviet Union and the United Nations sat on the sidelines with a different agenda; it was to pressurise the former three countries to withdraw.

The first Suez raids by both Valiants and Canberras from Malta, and Canberras from Cyprus, were carried out on Wednesday 31st October. Raids by both types of aircraft continued for six days, up until 5th November. Valiants from 138 Squadron (eight aircraft), 148 Squadron (six aircraft), 207 Squadron (six aircraft) and 214 Squadron (four aircraft) along with Canberras of 12 Squadron and 109 Squadron all operated from Malta and were tasked to bomb the Egyptian targets while other Canberras detached from the UK to Cyprus operated as visual markers for the bombers.

The first night of the operation, from the perspective of the Valiant aircraft, consisted of six Valiants of 138 Squadron detailed to bomb Cairo West airfield, five Valiants of 148 Squadron and one from 214 Squadron to bomb Almaza airfield, while five Valiants from 207 Squadron bombed Kabrit airfield, El Agami and Huckstep Barracks. Finally, two Valiants from 207 Squadron and two from 214 Squadron attacked Abu Sueir airfield. Unfortunately for 138 Squadron they were recalled, whilst airborne, from attacking Cairo West airfield as the Americans were still evacuating their people. However, over the following nights Valiants did attack Cairo West along with Fayid airfield, Kasfareet airfield, El Adem and again Huckstep Barracks. Throughout the detachment altogether the four squadrons of Valiants completed a total of forty-nine sorties.

The RAF's part in the short 'six-day war' was called Operation Musketeer. Some of the RAF, namely aircrew and ground crew, who took part, or associated in some way, relate their stories here.

John Foot, nav radar 138 Squadron Valiants RAF Wittering:

I joined the RAF on 5th February 1950 and later under the Empire Air Training Scheme I was sent overseas for aircrew navigation training flying Anson 2s at 3

Air Navigation School (ANS) RAF Thornhill, Gwelo Southern Rhodesia. In 1951 I returned to the UK, and RAF Swinderby, as a sergeant navigator where I attended 201 Advanced Flying School (AFS) flying Wellingtons. This was followed by 230 OCU flying Lincoln B2s and thence on to 61 Squadron at RAF Waddington. A posting to 97 Squadron at RAF Hemswell still flying Lincolns, interrupted by a short spell

Jim Catlin's crew return to RAF Witthering, 7th November 1956, after operations against Cairo West (recall), Cairo West again and Huckstep Barracks. Crew from left to right: nav plotter Fg Off Jim Eyre, AEOp Sgt James Robertson (Robbie), nav radar Fg Off John Foot, co-pilot Fg Off Geoff Rushforth, captain Plt Off James 'Jim' Catlin DFC, crew chief Chief Tech Ricky Salkin. (*John Foot*)

at Officer Training Unit (OTU) RAF Spitalgate, saw me return to the squadron commissioned as a pilot officer. Later a posting to the V Force, and to Valiants, saw me attending the long NBS course at RAF Lindholme and RAF Finningley arriving on 232 OCU at RAF Gaydon in May and finally on 138 Squadron RAF Wittering in August 1956 — just three months before the start of Suez.

I was the nav radar on the Catlin crew and with only three months on the squadron under my belt was detached to RAF Luqa for the Suez crisis. Although Jim Catlin[5], our captain, was only a pilot officer he was a very experienced pilot. He'd been on Lancaster Pathfinders during the war and was awarded a DFC. The reason for his current junior rank was only because he'd left the RAF and later re-enlisted after the war. As a crew we suffered having a so-called 'junior' pilot as we were given the oldest aircraft on the squadron, namely WP220. However, unknown to most it had the best H2S radar screen. Unfortunately, this was discovered much later by our CO who deciding that he would be one of the crews on the Strategic Air Command (SAC) Bombing and Navigation Competition at Pinecastle, Florida the next year, 1957, took our aircraft for his own!

We flew out to RAF Luqa on 26th October in 'old' WP220. On landing it was clear to see that the airfield was full of aircraft and full of detachment personnel on top of the normal Luqa establishment. Accommodation was tight and the officers' mess was full. However, luckily we were all accommodated in the visitors' mess.

My initial memory of our detachment was of all of us aircrew assembled in

---

[5]   Jim Catlin went on to fly Victors and was co-pilot in a Blue Steel Victor which went into a spin and recovered.

Catlin's aircraft until it was 'pinched' by the CO.
(*via millionmonkeytheater.com*)

the Operations Wing briefing room ready to be told what we were about to do. At the first briefing for the target, Cairo West airfield, we were told to bomb from 50,000ft. At this point a couple of the WWII pilots who had learned their trade over the Ruhr and 'The Big City', Berlin, asked if the aircraft could be coaxed any higher and could we use the gain in height? They were quickly told by a bumptious little fart, who shall remain nameless, that we were to bomb at 50,000ft[6] because he said so! Sanity prevailed when the Bomber Wing detachment commander told us to use whatever altitude we could reach. Another idiotic thing that we heard at the briefing was that the operation was to be called Operation Illegal which I thought was hilarious and my guffaw wasn't taken too well by the senior people present. However, later seeing their rather comical effort for an operation's name was a non-starter, it was changed to Operation Musketeer.

I carried out three sorties during my stay. Although the Valiant could carry 21 x 1,000lb in the bomb bay, we only carried 12 x 1,000lb bombs on each raid. The first night our target was Cairo West airfield. It was some time into the flight, on the way to the target, when we received a recall message that told us not to drop our bombs but to turn back. Apparently, we understood later, American nationals in the Cairo area were crowding to get out of the line of fire and thus still being evacuated at the time of our raid. We thought that this was a spoof message and continued flying towards Cairo. Shortly afterwards our squadron commander, Wg Cdr Rupert Oakley, who was leading our 138 Squadron aircraft was told that it was not an erroneous message. This was relayed verbally by a member of the operational planning team in the air traffic control (ATC) tower at Luqa. It was Gp Capt Woodroffe, our station commander at RAF Wittering, who got on the air and said "Rupert you've got to turn back". Recognising his voice OC 138 turned back and we as a squadron didn't drop our bombs. We landed back at Luqa with our load still on board.

Wg Cdr Oakley must have been hugely disappointed with the recall as he had to forgo the prestige of being the first to drop bombs from a Valiant in anger. It fell to 148 Squadron, commanded by Wg Cdr Brian Burnett, to receive the kudos as his squadron was also attacking that night, but on other targets. **John Foot** continues:

---

[6]   Valiants with bomb load couldn't reach 50,000ft.

John Foot's log book showing the Cairo West recall, 31st October, 1956.
(*John Foot*)

The second and third nights we were on operations again. This time, on 1st November our target was again Cairo West airfield with the same bomb load followed the next night by Huckstep Barracks. Each time, on our bombing runs, Jim Eyre, my nav plotter, was on his belly, doing the business lying in the nose using the T4 bombsight dropping the bombs visually while I backed him up with the NBS and following the target on the H2S radar screen. On the Cairo West attack the bombs were set for impact fusing. However, for the night of the Huckstep Barracks attack we'd been briefed that there was a lot of tented accommodation with the Egyptians under canvas. Therefore, to make life really uncomfortable for the enemy, and to get the maximum effect, it was decided to drop the stick of 1,000 pounders with airburst fusing set to go off at 1,000ft. This was our last operation and thus completed two nights of successful bombing.

After we landed and taxied in there was great consternation when we raised the deflector plate in dispersal for the crew chief to check inside the bomb bay as he found that we had a hang-up on one of the carriers. Luckily there was no mishap and the bomb problem was sorted out.

As to any opposition during our attacks we didn't see any. Although I did hear that one Valiant crew, not from our squadron, experienced cannon fire from an Egyptian aircraft.

Sqn Ldr Ware of 148 Squadron was the pilot in question and the fire was from an Egyptian air force Gloster Meteor NF13 that they easily evaded and out-manoeuvred. **John Foot** again:

After carrying out raids for three nights in succession the next day our detachment commander asked the station commander if we could have some of his men to help drive MT and to tow bomb trolleys. However, to our amazement despite all the chaos, the war going on and us working flat out, the station commander refused our request on the grounds that his men couldn't be spared as it was a station sports afternoon!

Whilst we were at Luqa the aircraft on the dispersals were guarded by the army, some of whom had never in their life been anywhere near an aircraft. Clearly one of the guards was so bored that he whiled away the hours on duty using his bayonet to engrave his name, rank and number on the black dielectric fairing that covered the radar scanner on the nose of the Valiant. Not the most sensible thing to do!

As already mentioned, regarding the sports afternoon episode, there was a certain lack of co-operation that we suffered from RAF Luqa personnel. Therefore, after the Suez 'hostilities' had ceased every man jack of them wanted to get in our aircraft or even be taken for joy rides. At this point our Bomber Wing detachment commander decided that it was payback time. Orders were given that no-one, but NO-ONE, was allowed to get into a Valiant unless he was a tradesman servicing something or a five-man crew going to fly it. Therefore, no RAF personnel from Luqa got aboard.

On the other hand, our friends and allies in the Royal Navy had provided a screen of submarines around the Nile Delta to fish us out of the water should that have proven necessary. So on every air test or sortie afterwards we were accompanied by the Senior Service. They, in their turn took us aircrew for voyages in their boats. I remember going to sea in HMS *Totem*, standing in the conning tower and listening to the crew's captain saying "What's for breakfast Number One?" whilst at the salute for leaving Grand Harbour, Valetta.

Finally, with the detachment at an end we returned to Wittering on 7th November. At this point another bizarre situation arose. Whilst on operations at Luqa it had been decided, sensibly enough back at Wittering, that we needed self-protection in the event of us baling out and having to come face to face with the enemy. Therefore, during our raids each of us aircrew wore an RAF issue webbing belt and holster containing a Webley .38 revolver worn over our flying suits. When we eventually reported to the armoury at Wittering to hand in our revolvers and ammunition it became a farce. In true tradition of RAF rules and regulations and sometimes idiocy, the armoury refused to accept the dozens of rounds we'd been issued before departure. So returning to the crew room looking like gunslingers every spare minute, when not involved with any other tasks, we had no alternative but to depart for the short range and run shooting competitions for weeks afterward. Anyway it has to be said that it was good fun.

Despite Gp Capt Woodroffe's war record, for which he was highly decorated, his role in Suez and his quick thinking in averting disaster on the first Valiant raids on 31st October, fate was to deal him an unlucky hand just one year later. While training four Valiant crews for the 1957 SAC Bombing and Navigation at Pinecastle Air Force Base (AFB), Florida he was killed in a flying accident, with the base commander and two other United States Air Force (USAF) senior officers, during a practice demonstration in a B-47.

The following is extracted from Philip Goodall's internet story *'Over Fifty Years Ago I Bombed Egypt'*. **Philip Goodall** co-pilot 138 Squadron Valiants, RAF Wittering:

In July 1956 Nasser nationalised the Suez Canal and in a matter of days our pattern of flying changed. If the Valiants were to be used in a war, the crews had to be trained to drop conventional bombs. As the new radar bombing system was still being developed and was not yet fitted to most aircraft, the Valiants obviously had to be provided with a bomb-aiming capability, thus they were immediately equipped with a visual bombsight similar to that used in the last war. High altitude visual bombing became the order of the day dropping practice bombs on every available bombing range in the UK.

In August 1956 the routine of my crew was suddenly changed when my skipper, Squadron Leader Bob Wilson, was tasked to establish that Valiants could operate from RAF Luqa in Malta. We flew out to Malta and carried out a full load night take-off. This operation highlighted

Bombing of Egyptian Airfields

A Press Release was issued by Air Headquarters, Malta, last night.

The statement said that three attacks on military airfields in Egypt were carried out on Wednesday by the Bomber Wing commanded by Group Captain L.M. Hodges, D.S.O., O.B.E., D.F.C., under the control of the Air Task Force Commander, Air Marshal D. H. F. Barnett.

These operations carried out by Valiant four-jet and Canberra twin-jet bombers were night attacks and bombing was done on visual markers. All bombs fell in the target areas. Weather conditions were good and there was only a small amount of cloud in one target area. All the aircraft employed reached the target areas successfully and returned undamaged.

Meagre Opposition

The opposition from the ground was meagre to moderate, the anti-aircraft fire coming far short of the main bomber force and no Heavy A.A. fire was seen. One encounter with a night fighter Meteor NF13 was re-ported but the Valiant attacked took swift evasive action and sustained no damage. Pilots reported that they had no difficulty in picking up Cairo and Alexandria.

Three Airfields

The three military airfields attacked by this wing were Almaza near Heliopolis, Kabrit on the shores of the Great Bitter Lake, and Abu Sueir, west of Ismailia.

The first attack laid on by Group Captain Hodges was carried out by a Valiant Squadron commanded by Wing Commander W. J. Burnett, D.S.O., D.F.C., A.F.C. Describing the operation Wing Cdr. Burnett said: "Our target was the airfield at Almaza between Cairo and Heliopolis. We set out on Wednesday evening and flew uneventfully on the 800-mile course until we came over the target at 8 p.m. We had picked out Alexandria on the way in and although I saw no street lighting in Cairo we could make out the whole lay-out of the capital."

2 - 11 - 56

Local newspaper 2nd November 1956. (*Roy Monk*)

two serious problems. Luqa was the only airfield with a suitable runway from which the Valiants could operate and none of the 1,000lb bombs in Malta were compatible with the Valiant. Fortunately the Canberra used a similar weapon, so the Canberra crews unexpectedly found themselves ferrying bombs out to Malta.

By October 1956, there were only four Valiant bomber squadrons, but some of the squadrons were still being equipped with new aircraft, which meant that the total force available was just twenty-four aircraft. There was frenzied activity on 138 Squadron as we made up a third of the total force. On the 19th we flew out to Malta. The day after our arrival we were all briefed at Air Operations and I recall my surprise at seeing a large map of Israel on one wall and a map of Egypt on the other, which made us all wonder who the enemy was to be.

As I have already explained, bombing with these new aircraft presented a problem as only a few of the Valiants were fitted with the new radar equipment, thus it was decided to use techniques developed during World War II. A radar-equipped Valiant would lead each attack and drop a red proximity marker on the selected target. The Canberras operating out of Cyprus would fly at low level using the light from the proximity markers to identify the actual target and then drop green target markers to be used as the aiming point for the bomber force. The first Valiant in each attack would drop the proximity marker and then orbit at high level to lead the bomber force to drop bombs visually on the green marker. Very obviously not the most sophisticated or accurate means of dropping bombs from 40,000ft and above.

There were two other aspects which surprised us. We were all issued with revolvers, as a means of protection in the event that we were shot down. We were also given a British government promissory note which offered to pay a vast sum to anyone rescuing the holder, the aptly named 'gooly chit'! Unfortunately I cannot remember the financial offer but I know they were assiduously collected after each sortie. I suppose the Air Staff were worried that if we discovered we were only worth £100, we might go on strike.

On 29th October Israel invaded Egypt and on 30th October, Britain and France threatened to invade unless Israel withdrew from the Canal Zone. On 31st October we bombed Egypt.

My captain was leading the attack on the Egyptian air force base of Abu Sueir from an altitude of 42,000ft. As we approached Egypt all looked peaceful. The lights in the towns and cities were glimmering below. Our eyes were searching the skies for any signs of enemy aircraft. The radar operator identified the target; after a steady run, "Target Indicator away". We turned and prepared to make our attack with live bombs. The sky was illuminated by our red proximity marker. Shortly afterwards the Canberra pilot came on the radio, "Identified the target". We waited for what seemed like hours but must have been minutes, then the sky was lit up again but this time by the green marker, followed by instructions from the Canberra pilot, "Bomber force bomb on the green marker". By this time we were running in for our second attack; "Right, steady, right, steady, steady, bomb doors open, steady, bombs away." All bombs dropped and we turned back towards Malta followed in turn by the other aircraft in our raid. All appeared to have worked according to plan and the entire force returned to Luqa some five and a half hours after take-off.

The next day we all anticipated great publicity and applause for the new high level air aces, only to learn of the discord and argument at home. Certainly the news had a deflationary effect on the whole force with the realisation that our operations were the cause of serious disharmony. Nevertheless we had a task to complete. Attacks continued for the next five days on a variety of targets, specifically seven airfields, two military barracks, a naval repair depot and a railway marshalling yard. All the Valiants returned without damage, though ack

ack fire was evident at certain targets.

Following the six days of operations, it was evident that the politicians rather than the military were fighting, so we made a quiet return to the UK on 7th November, less than three weeks after we left home. Not surprisingly, I suppose, we were met by the press. A Scottish friend was asked how he passed his time in Malta and assured the reporter that Egyptian callisthenics was his pastime and duly earned national publicity. The RAF smiled, as in our parlance, Egyptian PT was sleeping! My crew was met by Pathé News with the result that my mother sat three times through the films in order to see the news and I am quite certain that the whole cinema would have known that it was her son.

The report on Operation Musketeer, the Suez War, details the RAF participation. 24 Valiants operated out of Luqa with a further 29 Canberras based at Luqa and Halfar in Malta and 59 Canberras operating from Nicosia in Cyprus. In six days of operations a total of 259 sorties were flown with 942 tons of bombs dropped.

> **John Roberts**, navigator 21 Squadron Canberras RAF Waddington, was involved in the preparation for the Suez adventure:

Although I wasn't actually involved in Suez we, on our squadron, found it strange that, out of the blue, on the run up to the war, we were suddenly ordered to start bombing from 40,000ft. As we normally bombed from 15,000ft, as a matter of course, this diversion from our normal practice didn't seem to make sense and nobody gave us the reason for the change in tactics. Although I was posted just ten days before the first raids, 21 Squadron did take part in Suez. I also knew at the time that there was a shortage of 1,000lb bombs in the Middle East. Some of the Canberras were used to ferry them out in their bomb bays, six bombs at a time. In order to load the bombs on a Canberra the procedure was to jack up the aircraft. This, of course, took time and so at the receiving end the armourers just pressed the bomb tit and somehow got the bombs to land on a bomb trolley positioned underneath the aircraft!

> Following Nasser's decision, on 26th July, to nationalise the Suez Canal Company, plans were set in motion for Canberras to ferry out 1,000lb bombs to RAF Luqa. Ferrying started at the end of July and was known as Exercise Accumulate.

> **Roy Monk** was Corporal Engine Fitter 214 Squadron Valiants RAF Marham. His main story is in Chapter Seven but he was also involved with Suez:

On 20th October, along with others on the squadron at RAF Marham, I was told by my flight sergeant to pack up my things. We were all going to Malta. The next day we found themselves on board a Shackleton heading for RAF Luqa.

My most vivid memory of that particular flight was that we experienced a horrendous storm over Paris, the aircraft was struck by lightning and aerials ripped off. The aircrew had no option but to declare an emergency and we diverted to Marseille. All us passengers left the aircraft and crew behind and carried on by road along the coast a short way to the French air force airfield of Istres. Our Shackleton arrived the next day and we continued our flight to Luqa. On 26th two of our Valiant aircraft arrived at Luqa namely: WZ377 and WZ393. On 29th October we were told that we were all confined to camp [undoubtedly for security reasons]. On the 30th our second, and last, pair of aircraft arrived namely: WZ395 and WZ397. All we knew at this point was that our aircraft and other Valiants and Canberras on the airfield were going to bomb targets in Egypt. We weren't told the targets.

I watched the take-offs of both Valiants and Canberras each day and night when I was on shift. One particular night I recall was when an aircraft came back landing with two hang-ups. They were discovered by the armourers when the bomb doors were opened just as we were about to re-fuel the aircraft. When we heard about the situation we disappeared 'at a high rate of knots' until the armourers sorted it out. On the fifth night of operations, 4th November, we were told that it was to be the last of the attacks by the Valiants. At the same time were also informed that during a previous raid one of the Valiants from 148 Squadron was attacked by an Egyptian air force Meteor NF13. However, it was successfully evaded, and out-manoeuvred, by the Valiant. On 7th November all the Valiants from the other three squadrons returned to the UK, leaving us behind at Luqa. Our Valiants, WZ377, WZ379 (which had replaced WZ393), WZ395 and WZ397, followed at a later date. On 12th December the rest of our squadron detachment returned home in a Hastings courtesy of Transport Command.

So ended the UK, the RAF and the Valiant involvement in the Suez debacle.

Chapter Five

RAF REGIMENT SUEZ –
GUARDING THE VALIANTS

This first-hand account of the Suez operation is unique as it comes not from aircrew nor ground crew but from a RAF Regiment NCO. **Terry Gladwell** spent the campaign guarding the Valiants at Malta and splendidly presents an interesting view of the operation.

In August 1956 the RAF Regiment Squadron of which I was a part was en route from Felixstowe, Suffolk to Watchet in Somerset to fire our Bofors guns on the range there. We never reached Watchet, for the Suez emergency intervened and we were turned around at Bicester where we had overnighted. Back we went to Felixstowe and from being a light ack ack squadron we were transmogrified into a field squadron, quite a different arrangement. There was much coming and going of personnel, and a whole new structure emerged. Days later the now reconstituted 63 Field Squadron headed west, still with the Bofors guns, but at Innsworth, near Gloucester which was then the RAF's overseas transit station, the guns were parked up, the lorries handed in, as it were; we all got tropical kit, an update of jabs, and a lot of assault-course training in our new, more basic infantry lifestyle.

Soon we were on our way to the Middle East (we thought) via Clyffe Pypard and Lyneham, Wiltshire. An advance party from No 1 Flight had travelled this route some days previously to wherever 'Their Airships' decreed. Armed to the teeth, carrying loaded weapons and expecting to be delivered into the middle of a firefight, the rest of 63 Squadron boarded a whole bevy of Handley Page Hastings, the RAF's ultimate troop-carrying aircraft of that time. (Personally, I thought the service's fixation with rearward-facing seats – a safety measure, we were told – was a bit curious and off-putting, never having been fond of travelling with one's back to the engine.) We took off and headed south-east towards our first stop, at Luqa, Malta.

As it turned out, it was the only stop. We deplaned and were taken to RAF Safi, a small station about three miles from Luqa. Safi was a workshop facility and storage unit. Our advance party had erected tents for the squadron there. Next morning we learnt of our future duties. The island of Malta was to act as a gigantic aircraft carrier, with a planned offensive against Egypt being mounted from its flight-deck, or runways. The spearhead of this attack would consist of the Valiants of the V bomber force, which was then the RAF's 'Sunday Punch'.

Our job was to safeguard these bombers, each one rumoured to cost around £1,000,000, on their dispersal areas at Luqa. At the moment there were only four of these behemoths at the airfield, but more were expected daily, and security was to be absolutely 100% around them whilst they were on the ground at Luqa.

The perceived danger was that Egyptian commandos, or even Maltese Egyptian sympathisers, might seek to interfere with the aircraft in some way, and we were to prevent this.

Our squadron had a strength of around 150 men, including RAF tradesmen such as stores, MT, medical, armoury and clerks. The regiment establishment for a field squadron was three rifle flights, a 3-inch mortar flight and a small headquarters flight. We had perhaps 120 combat troops available for this security task. It didn't help much that we were rostered immediately, numbers 2 and 3 Flights doing 24 hours on/24 hours off duty at Luqa. Jammy No 1 Flight, which had been the advance party, it will be recalled, were again sent elsewhere, this time to RAF Idris in Libya. They airlifted in Shackletons of 38 Squadron from Luqa, and we never saw them again for four months.

Those of us on No 2 and No 3 Flights alternated daily doing the guard duties. At first, it was quite easy but more Valiants arrived at Luqa and we got pretty stretched trying to cover them. 'Malta Dog' didn't help, either; this was a nasty type of diarrhoea, caused by eating vegetables which had been grown on the island. The only treatment was copious libations of kaolin and morphine, a fawn-coloured medicine tasting of toothpaste and liquorice which somehow made you better after a few days burping!

Luqa was very busy. The offensive was hotting up, and there were a lot of extra aircraft not only based there, but visiting in relation to the Suez Crisis, as it later became known. Two Squadrons of Shackletons, 37 and 38 were long-term residents; there were Hastings, Valettas, Canberras, Ansons, Venom fighter-bombers passing through and even some of the first Hunters called in. More exotic transports came and went; an early Blackburn Beverley, a Comet of 216 Squadron, and several RAF Lockheed Neptunes leased from the Americans, now being withdrawn but which were ferrying through Malta en route to the UK for disposal. From the French L'Armée de l'Air we saw Corses, Bellatrix, Flamants and even a Noratlas. There were civil flights with BEA's Ambassadors and Dakotas, and BOAC's Argonauts, Britannias and Yorks. At nearby Takali there was a squadron of RAF Meteor FR 9s, (208) and at RNAS Hal Far, three miles the other way, masses of Canberras from several Bomber Command squadrons had assembled for the coming campaign.

The Valiant dispersals were of the large loop design. Each contained just one aircraft, and was floodlit with sodium lamps at night. Our sentries were each allocated one dispersal and worked a 'two hours on/four hours off' shift pattern for the twenty-four hours of our duty. They weren't given any ammunition, which was rather negative thinking, but just a rifle, bayonet, a torch and a whistle. (I don't know why this was, but it didn't actually inspire anybody.) Plain and straightforward as the task may have seemed, there were incidents a-plenty.

There was, for a start, 'The day of the leaking Valiant'. For much of the year Malta is hot, dry and fairly windswept. In late August and on throughout autumn, however, the rainy season begins, and when it rains, it really chucks it down. With the torrents which had teemed unceasingly all the preceding night having at last eased off, the sun came out warmly and everywhere the temperature rose by leaps and bounds. A whistle sounded on one of the dispersals two hundred yards away from the guard tent, and with one of my corporals, Johnny Rhea, I galloped over to see what was wrong. Gunner Jim Dunks was on this particular post and he was alarmed, to say the least. From two downward-pointing spouts, one at each wingtip and about the size of household rain pipes, liquid was running out in a steady stream. I held a hand in the flow to catch a drop of this stuff on my fingers, and smelt it. It was jet engine fuel, or something like it, for it had a strong paraffin odour. We had a hurried consultation on what to do next; I thought I could understand why this was happening, but there was an awful lot of this fluid gushing out.

Rushing back to the tent, I phoned for the aircraft servicing party to come and have a look for themselves. This took a while, because it was a Sunday morning, fairly early and with not much else happening anywhere except on 63 Squadron. Finally a Land Rover drew up, disgorging three or four airmen with a chief technician in charge. The fluid from the wings had now ceased to a trickle, but there was a large puddle of the stuff on the cement as proof, if proof were needed.

The crew chief explained to us that the aircraft's tanks had been topped up the previous day, in readiness for a flight later this afternoon. The sudden warmth of the sun after the coolness of the long rain had made the fuel expand within the tanks which were installed inside the Valiant's wing structure. The overflow caused by the expanding volume had travelled along the plumbing to the dump spouts which were carrying out their designed function. Usually, it was said, this sort of thing happened when the aircraft in flight passed through different height and pressure levels, or layers of warmer air. He was happy to see that we were 'clued up' and not afraid to send for them if strange things happened with the aircraft. The fuel which was on the ground would soon evaporate harmlessly and as the level of the tanks stabilised, the flow would eventually cease completely.

The servicing party went away and I logged the occurrence in 'The Book', just in case they decided to make out a report and I hadn't, which would look a bit odd. It was the old service maxim, 'the desire to be covered'. As long as you had something in writing, you were fairly safe from repercussions.

Then there was 'The day of the Strange Person'. Privy as we were to all the refinements and accomplishments involved in preparation for the proposed bombing offensive, we had been continually reminded and preached to concerning the provisions of The Official Secrets Act. In particular, it was stressed to us that we must abide by the bit prohibiting 'any sketch, plan, model, photograph, article, note or document', all of which were very strongly forbidden unless with the benefit of official sanction. We were told that as we were on 'active service' for the duration of the conflict, however long that took, even our mail home would be censored by an officer of the squadron. So we were all pretty twitchy about possible espionage taking place.

Therefore it was with some surprise and not a little worry that Cpl Gordon Pettifer and myself observed a person, dressed in shabby old clothes, walk up the peritrack from the sharp end of the airfield one afternoon. When some three hundred yards or so from the nearest Valiant dispersal, this person stopped, set up an easel and began painting a likeness of the aircraft. This just had to be investigated, for our own sake, in case the flight commander, with whom I was not exactly flavour of the month, came around on one of his sudden swoops.

Taking an airman with me, I made my way across to where this artist was busily sketching away. Without being rude, for on getting nearer I could see it was an elderly female, I asked if she had permission to do this. "Oh, yes my dear," she said "I've got lots of bits of paper...would you like to see them?" With that she delved into a voluminous shoulder-bag and brought forth a great deal of permissions, issued by the Air Ministry, the Admiralty, the War Office, even the NATO High Command, and several other such august bodies. Hastily daring to examine these documents, personally signed by very very important persons, such as admirals, air marshals and generals to name but a few, I quickly handed them back, for it seemed they were so red-hot it was burning my fingers. I explained my motives for checking-up on her, and with very good grace she said she understood completely.

This was the first and only time I ever accosted an official war artist in the course of duty, and we parted amicably, for the light was beginning to fade. Her efforts were not at all bad; she had drawn superbly in the short time that she had been there. My only regrets are that I never saw the finished painting either in reality or reproduced in print.

Then, and it could only have happened to a 3 Flight guard, there was 'The day of the smoking bomb dump'. The rain, which had fallen in sheets for the past three days and nights having at last ceased, the sun emerged victorious from behind the cloud line, and everything in the garden was lovely. Our garden, from the guard tent, being the bomb dump, where we had a marquee and other facilities set up on a knoll at the far end to accommodate us during our 24-hour shifts.

The offensive against Egypt was obviously expected to be a long and hard one, for this quarry-like repository of aircraft hardware had, during the past month or so, received much in the way of additional lethality, stack upon stack of which was crammed inside. There was no roof, nor any sort of building in the dump; it was just a large excavation chiselled out of the ground and rock, with a little tarred road running through the centre.

On either side of this road were the bombs, in heaps; small bombs, big bombs, flares, target markers, incendiaries, high explosive, armour-piercing, general purpose, special purpose, depth charges...in fact every sort of air-delivered weaponry imaginable. They were all in tidy piles according to their purpose, just like bricks or sewer pipes stacked in a builder's yard, and there were literally thousands of them.

We had just taken over from 2 Flight guard, who were now back at Safi getting their heads down after a long day and night. As the weeks progressed with no relief from the 'day on, day off' routine, we were all finding that sleep was more important to us than anything else.

Johnny Rhea, Gordon Pettifer and myself were sitting outside in the sun, overlooking the bomb dump. Nothing much was happening, it was an hour before we needed to change the sentries, and it was such a pleasant morning after all the downpour that it seemed a pity to stay in the tent, making out changeover rosters or suchlike, or waiting for the telephone to ring. Suddenly Gordon spoke:

"That's strange. I thought I saw smoke down there!"

"What, smoke in a bomb dump? You must be feeling the strain. Go and have a lie down", Johnny and I both chaffed him.

"No, I'm sure I did. I'm going to have a look!"

We both went down the path with Gordon to the floor level of the dump. He led us fifty yards through the high stacks of incendiaries and target markers. Sure enough, there was smoke, coming from a pile of white-painted 25lb practice bombs. We didn't wait to see any more! As they used to write in novels, with one bound we were free, getting back up on top of the knoll where the guard tent, and more importantly, the telephone was.

A cold sweat formed on my forehead as I dialled the number for the station armoury. After ages, a disinterested voice took the call. "This is the regiment guard-post at the bomb dump", said I. "There's a pile of bombs smoking in the dump. You'd better get here fast and bring the fire brigade with you!" He still didn't seem unduly perturbed; perhaps he hadn't believed me. It was way past April Fool's Day, so I repeated the message again, adding that I was deadly serious...what a term to use!

"Righto, Sarge...we'll send somebody out to have a look shortly", was all I could get by way of satisfaction. 'Shortly!' I thought, 'How very matter of fact... does this sort of thing happen often?'

Replacing the handset of the telephone on its cradle, I looked at my colleagues as I did so. It was important not to show my inner feelings at this time, which were in absolute turmoil, but to put a brave face on it; a sort of 'Steady the Buffs' attitude. Understandably, they too were eager to know what was being done. So far none of the airmen in the main tent next door knew what was going on, but they would soon know if things started to go OFF! Not that it would help much for if this lot which we were in the middle of exploded, most of Malta would go with it, so there was nowhere to run and hide, even assuming we had time. Every second weighed a ton!

A leisurely Land Rover turned in at the far end of the dump and stopped. A couple of languid-looking figures got out and walked towards us. I plucked up courage from somewhere and went to meet them. It was a sergeant and an airman from the armoury, and I led them to where this wispy tendril of smoke was still rising from the 25 pounders. "Oh, that's all right", said the sergeant in what I thought was a condescending tone. "Some of the rain must have got inside the casing and caused a chemical reaction with the filling. There's no danger of any explosion."

So spake the voice of learned technology, and I felt a bit of a chump for having leapt at the obvious conclusion, and said I hoped they would understand my anxiety. They assured me that there was no reason for alarm, and shook the water off the rest of the bombs in the pile, taking two or three badly affected ones away with them. It wasn't even smoke we had seen, but a kind of gaseous vapour of chemical reaction. All the same, it had given us a nasty few minutes...talk about sitting on top of a time bomb!

To guard the Valiants sentries were needed for 24 hours a day and so we had to get help. Our bosses, far above squadron level, were compelled to ask the army for assistance in this matter. Help was forthcoming in the shape of a company from one of the West Country light infantry regiments, who were given a sector of Luqa airfield to cover, far away from where we were operating.

All was well for a few days until one of the 'brown jobs', possibly bored with walking round and round the big silver bird for two hours at a time, spiced his day up by carving 'Malta 1956' and his name on the fibreglass di-electric panel which formed the lower part of a Valiant's nose...using his bayonet or some other gouging implement.

There was a huge furore when this was discovered. Because of damage to the pressurised part of the nose, the Valiant had to be flown back to the UK straightaway and at reduced height, for repair. I don't know what happened to the soldier who did the etching and kindly provided his own signature to prove it; he's probably still doing jankers today.

The attacks on Egypt are now part of RAF history and have been fully documented elsewhere. The Valiants from Malta, representing detachments of several different squadrons, were responsible for the high-level bombing of the Egyptian airfields. Canberras from Malta and from Cyprus looked after the medium and low level raids, supplemented by Venoms and Hunters, also from Cyprus. Naval aircraft, both land and carrier-based, were likewise involved.

At one time, there were twenty Valiants parked on Luqa airfield, and it was quite a daunting thought to be guard commander in charge of £24,000,000 worth of aircraft, definitely keeping us on the hop, at three separate locations around the base. During this intense period, 63 Squadron was operating one flight of guards by day and two flights by night, at three separate locations on the airfield.

Before the campaign against Egypt was abruptly brought to a halt, following international pressure by both the US and Soviet Russia, around the clock missions from Malta had meant an almost constant activity by the Valiants, which made our job more complex and difficult. No longer was there time to carry out the formal changeover ceremony whenever one flight replaced another, as had been the case at the start of the twenty-four-hour guards...obviously someone on the squadron had thought we were at Buckingham Palace or the Tower of London, with all the formality that such a handover generated. Instead, we merely arrived on site, and took over as a matter of course; much more quickly and with no archaic bull dust surrounding the event. Everybody was aware that there was something 'big' going on and knew what was required of them by habit. Nevertheless, with an abundance of 'scrambled egg and fruit salad' taking close interest in the more glamorous operational side of the RAF, we couldn't afford to get slipshod or complacent, however many times we had done all this before.

On more than one occasion as guard commander, I had to politely explain to the 'flying' side just why it was that everyone who approached one of these silver or white monsters was challenged at bayonet point by a sentry. They hardly saw it as necessary, and in a way I could understand their point of view, being diplomatic about it, for it was 'their' aircraft we were looking after. The Valiants hadn't just been parked at Luqa to give 63 Squadron guarding practice. It worked both ways though. One of 2 Flight's sentries had been charged on an aircraft captain's order; through not challenging the aircrew or examining identity papers. After that, anybody and everybody got challenged and asked for documentation; from the aircrew right down to the bod who drove the refueller truck or the bomb tractor.

The Valiants themselves were marvellous beasts, beautifully made and of course, brand new in many instances. Squatting hugely on the dispersal pans, it was unbelievable that they could ever fly. But once in the air they were so graceful and elegant it was hard to see them as machines of destruction. There can have been fewer aircraft ever designed to deliver bombs with so fine and handsome an exterior. Most were in silver; some even in what might be described as an 'antique' silver, more of a grey, and there were a few in a kind of white livery with toned down markings. At the height of the campaign against Egypt there were aircraft from 7, 138, 148, 207, 214 and 543 Squadrons at Luqa. Not, it must be emphasised, the complete squadrons mentioned, but detachments from them all. We got on very well with 138 Squadron detachment, the first-ever Valiant unit, under the command of Wg Cdr R G Oakley. He encouraged an affinity with the regiment and did a lot to make our guard duties as comfortable as they could be in Nissen huts allocated to his dispersals. This was really something for by now the autumn weather had taken a turn for the worse and when it wasn't actually raining there were decidedly nippy winds sweeping across the airfield.

On Malta, apart from the usual sea-bird life at the coast, there is very little else in the way of feathered activity. There are numerous possible reasons for this paucity; the distance from mainland Europe and Africa; the lack of natural nesting sites with very few hedges, but stone walls instead; the shortage of grain crop production on the island... but as I firmly believe to be the main cause, the predilection of Maltese villagers to go out every weekend shooting at anything that flies within range of their ancient weaponry. At times when migrating birds call in to Malta for a quick rest, this slaughter heightens, and very few of the overnight stoppers make it any further. Most of the guns are filled with a mixture of tin-tacks and other scrap-iron to give a lethal scatter pattern when discharged at the ingredients of 'sparrow pie' or whatever the culinary outcome is called.

One of my sentries had been replaced at the Valiant he had guarded when his 'stag' was up. Soon after, the ground crew arrived to prepare the aircraft for flight and it was shortly after this that to coin a phrase, 'the muck hit the fan'. They discovered a sizeable hole in the aircraft's fuselage, just below the cockpit on the starboard side. As a result, this Valiant was now rather more than just unserviceable, it was definitely 'hors de combat' as far as the foreseeable future was concerned. By now the aircrew, suitably togged up and primed for their foray into the blue yonder, had also arrived. They were more than a bit miffed, naturally, to find out that they had wasted several hours getting kitted up and briefed for a 'no-go'.

The new sentry was called over by the aircraft captain and told to get me out there, on the spot, pronto, like as of yesterday, so to speak. The first I knew about it was when the new sentry came back to the guard tent a bit early, I thought, and not having been properly changed over or anything. I soon got the message and ran back with him to where both the air and ground crews were dancing up and down in high old dudgeon. With all this hold up, the other three Valiants from this particular dispersal area had lit their fires and steamed off Luqa for points unknown, leaving this wounded bird very much in the nest. A whole convoy of cars was charging along the peri-track; their arrival coinciding with mine. They disgorged quite a collection of important-looking brass, who went into a huddle with the aircraft captain; then trooped around the starboard side to study the new and unofficial modification to the fuselage. I was asked if the other sentry had heard or seen anything strike the aircraft. I replied truthfully that if he had, nothing had been reported to me when he came off shift. I was told to get my officer out there to take further steps and it was clear that something was going to happen. Johnny Rhea, clued up as ever, had already phoned for the flight commander to come from Safi.

Flying Officer Wilkinson and Flight Sergeant Melville were soon on scene and joined the crowd around the aircraft. I was sent back to the guard tent to locate the airman who had previously been the sentry. He was one of our better lads, but they were all a pretty good bunch anyway. He was quite bright. When questioned by the flight commander he said that whilst on his shift he had heard a sudden thump come from the aircraft but hadn't noticed anything odd. This wasn't all that unusual, the Valiant was a large aircraft, with acres of metal, which creaked and echoed all the time as the result of temperature changes, such as when the sun went behind a cloud, much the same as a tin roof on a shed does; at times quite noisily. There was some truth in what he said; the mass of metal had to expand and contract. He also said there had been so much banging coming from the nearest village (Birzebuggia) with the shotguns going off at birds, that one bang sounded just like another or an echo thereof.

He was told to wait outside, and I knew it boded ill for him when Chiefy Melville said to me, "That's it, John (he always called me John, whenever a rocketing was imminent)....you'll have to charge him for not reporting this damage to you." And so it was that the airman's name was inscribed on a Form 252 and in due course the lad came before the deputy squadron commander and got seven days jankers. It was perhaps slightly harsh, but there again, he ought to have been aware of the damage happening and thus mentioned it to me. There was no doubt it was caused by a local with a shotgun whether accidental or on purpose. If I had known I could have informed the ground crew and started the investigation off; by my not knowing, they had found it for themselves. You could follow their reasoning... 'just what are these Rockapes supposed to be doing? Looking after our aircraft or wandering about with everything switched-off?'

Much, much later in the time of Musketeer (a name we only heard of after it was all over), illness and wear and tear on our numbers had so reduced our efficiency to a level where we were unable to cover all the guard points needed, and consequently our commitment exceeded our strength. To make matters worse, there had been a steady escalation in the number of Valiants and other valuable and attractive aircraft.

When the offensive had been halted and the Valiants returned to the UK, it was obvious that our duty had gone with them. All the same, we were quite proud, in a way, to have been selected to guard them, and to be the first-ever RAF Regiment Squadron to be based on Malta.

One unusual anachronism with the Valiant sticks in my mind. On the nose of the aircraft a short pipe housed the socket for an extension intended for mid-air refuelling, whereby a special nozzle to connect with a fuel-delivery line drogue could be fitted. I suppose the diameter of this socket was about five inches. Because the nozzles were not at this time being used, the sockets were closed up with a bung-like stopper, which upon close examination turned out to be hardwood. It was possible to see the grain on these rounded bungs where the silver paint had worn off with the airflow and weather erosion. Fancy a million quid's worth of high technology relying on a bit of timber to keep the wind out!

Chapter Six

EARLY SQUADRON DAYS

Alan McDonald remembers the early days of the Valiant and remained with the aircraft right through its life until just before the main spar problem was 'discovered' when he went on to Vulcans with Blue Steel.

It was August 1955, a couple of months before my twenty-second birthday. I had been flying Canberras at Wittering on 76 Squadron for a couple of years when, out of the blue, I received a posting notice to go to Gaydon to become a co-pilot on 543 Squadron. At first I was not best pleased, as I was enjoying both the squadron and the Canberra.

The V Force was just starting up and the very first Valiants were arriving at Wittering. When my posting arrived it was to go on to the Valiant as a co-pilot; to be a captain you needed 2,000 hours first pilot and I had about 850 at the time; consequently nearly all the captains were ex-wartime flyers. After a while I got used to the idea and I was quite excited at the thought of getting in to a new force at the very beginning. However, my confidence in the aircraft received a severe blow. I was standing on the airfield at Wittering with my 'clearance chit' in my hand when a Valiant took off – it was such a new thing then that we all still stopped to watch – but when the aircraft got to a couple of hundred feet, it rolled to the right and crashed in a fireball about a mile off the end of the runway. I just stood there dazed, wondering what sort of aircraft this was going to be. The whole station went into quiet shock, it was a real tragedy to lose five of our own.

Fortunately, the cause was soon discovered – too late for that crew, but soon enough for the rest of us. It was a runaway aileron trim that was far too powerful for its purpose. In the event, the aircraft was one of the best, good to fly and kind to you if you made a mistake.

We moved to Gaydon in Warwickshire to do the conversion course, parking the caravan on a windy disused runway, and I got on with grappling with the intricacies of the Valiant's electrical system. Most people on the course were a lot older than me and there was a good sprinkling of DSOs and DFCs around. There was one captain, though, Harry Clerbaut, who was still a flying officer like me because he had been an NCO wartime Mosquito pilot, complete with a Belgian Croix de Guerre. He asked me to be his co-pilot as I was the only one who did not outrank him and so began a good working relationship. If I remember rightly, we were the only all flying officer crew, with Pete Hinchcliffe as nav plotter, Andy McCully as nav radar and Ernie Butcher as AEO. The Valiant was a big aeroplane and it took a bit of getting used to but it was good to fly and, at that time very modern and sophisticated. At the end of the course we were posted to RAF Wyton, near Huntingdon, to start up 543 Squadron whose role was photographic and radar reconnaissance. The whole course was made up of the second half of

138 Squadron and the first half of 543 Squadron.

The RAF was full of 'characters' in those days and Harry Clerbaut was certainly one of them. He said that I was no use to him unless I could do everything that he could so that I could fly the thing when necessary. So, he taught me well and often I did have to fly quite a bit of the sortie. In this respect I was very lucky – some co-pilots never got a landing. I got a very fair share of all the flying and if it was my turn to land, blowing a gale and pouring with rain I still got to do the landing. Harry was a great pilot, a good teacher and a pleasure to fly with.

On this squadron we had many overseas detachments and quite often spent our time flying very near to national boundaries that did not welcome the reconnaissance. We were frequently around the 'Med' and in the Arabian peninsula, especially from Aden and Malta.

October 1956 Operation 'Snowtrip' Detachment to Royal Canadian Air Force Base (RCAF) Namao, Alberta

The most significant detachment, and the longest, was to Canada. We went for three months in the autumn to see the terrain change from foliage to snow cover, came home for the odd month and then went back to see the thaw and,

finally we did a third, shorter, detachment in the summer to see what the radar pictures looked like in high summer. We were copying the terrain of Russia, for this was the Cold War era; by flying over northern Canada and going through the various seasons we could see what effect the change to snow and ice made to the bombing radar returns. In fact the results were quite dramatic and the trial was

From left, Pete Hinchcliffe, Alan McDonald, Fg Off Downer, Ernie Butcher and Harry Clerbaut.
(*Alan McDonald*)

well worth doing. It was a good trip for me too for I was the co-pilot on the Valiant, but also flew as captain on the two photo-reconnaissance Canberras that were also on the trial. It also started my long-term affection for Canada which has lasted right up to now. We flew way up into the Arctic regions, over miles and miles of nothing except frozen tundra, lakes and rivers. I have never seen so much hostile wilderness. The radar results we got in the air had to be verified against ground temperatures and snow depths and these results were obtained from Royal Canadian Mounted Police posts, logging stations and remote mining communities. Some of these we got by flying an old Beechcraft Expeditor into remote airstrips, mainly landing on rolled snow, and sometimes we would drive north up the icy Alaska Highway and pick up the results that way. We stayed in some very strange places!

I was married and on these long detachments the local overseas allowance was really helpful for we had to live in two different places but what was a bit of a swizz was that, being under twenty-five still, I was not getting the full amount of overseas allowances. In fact I was the only one on the whole detachment except a young pilot officer getting the lower amount and it was a source of irritation to me for it was the same on all our many overseas detachments! So my wife Eileen's work was vital to keep us both going. I spent about six months in Canada in all, coming home every couple of months to do major servicing on the aeroplane.

Unfortunately this also included Christmas 1956, for just as we were about to fly home in time for the holiday, all our tyres burst on landing and we ground to a halt wearing out the wheel hubs as we did so. I got the blame initially as I did the landing, but, fortunately the engineers found that the brakes were stuck on by a broken component. It did not alter the fact, though, that we had to wait until after Christmas for some spare wheels and tyres to be flown out from the UK. We were all made very welcome by the families on the base at Namao, however, it was not the same as being at home with Eileen.

In the end we got home on New Year's Eve just in time to go to the mess for the party, having flown two legs, Namao to Goose Bay, a quick refuel and then on to Wyton.

The crew at RAF Namao 1956/7. (*Alan McDonald*)

May 1959 saw the end of my co-pilot's tour and I was lucky enough to be one of the first 'boy' co-pilots to be given a captaincy on the Valiant. We took the caravan back to Gaydon while I did the first-pilot's course – back to the same windy runway site. I crewed up with a motley bunch, a co-pilot, Jimmy James, who had just finished flying fighters in Aden, a nav, Jim Carpenter from Canberras and two warrant officers as AEOp and nav radar, Geoff Baker and Wally Kerans. They were both ex-wartime aircrew and were very experienced and I often wonder what they thought of flying with a 'sprog' captain like me. As it turned out we all got on very well and I think we turned into a half-decent crew. We got a good squadron posting too, on to 214 Squadron at Marham, the very first flight-refuelling squadron in the RAF, given the task of introducing in-flight refuelling and all the planning and procedures that went with it. Our euphoria was short lived, though, for after a couple of weeks we were transferred on to 207 Squadron, which was a straight nuclear bombing outfit and this was not such an interesting job. Apparently they were one crew short for the start of a quick reaction alert (QRA) and we, as the junior crew, drew the short straw.

So we settled in to the job of becoming a 'combat' crew on 207 Squadron, but one bizarre incident happened to us in the process. We were getting ready to fly a practice atom bomb sortie – it was huge and weighed in at 10,000lb. As we were doing the pre-flight checks and Jimmy and I were just climbing up the ladder to get in, the aircraft almost leaped into the air as the bomb fell off! Jimmy fell back off the ladder on top of me and when we got up we rushed under the aircraft to see if everybody was all right. Wally Kerans, the nav radar had just walked away from the weapon as it fell and was very lucky, but it had trapped a corporal armourer under its giant tail fin and he was crying out in pain. We then did one of those things that you could not do under normal circumstances, we lifted the back end of that huge bomb and pulled him out. He had been trapped by the arm and it was obviously well broken, almost severed, so we piled him in the back of the ground crew's Land Rover and rushed him off to sick-quarters. There a very able young doctor strapped his arm together and got him as soon as possible to Ely hospital. Some six weeks later he was back on the job with almost full movement in his arm and hand. Needless to say we did not fly that sortie and the aircraft was soon surrounded by a big canvas fence to keep prying eyes away while a very thorough investigation took place. The reasons it fell off were quite complex and, needless to say, several different safety procedures were introduced. I must say that it shook us up a bit.

Then an incident happened that was very sad but benefited us as a crew. One of the crews from 214 Squadron took off in the middle of the night to do a flight refuelling exercise and crashed shortly after take-off – there were no survivors. This meant that suddenly the tanker squadron was short staffed and we were transferred back. They say that it is an ill wind that does nobody any good.

So, we started all over again to learn the black art of flight refuelling, both giving and receiving fuel. It was good flying and very enjoyable, although the 'prodding', as the receiving was known, was quite tricky at first until you got the hang of it. Eventually I could do it easily by day, but by night it was always a bit dodgy. The only lights on the drogue were three small bulbs, powered by bicycle-type dynamos – you could not have higher voltage at that point because a spark could ignite the fuel as you started on the transfer. Sometimes these lights worked but often they didn't – sometimes you had only one bulb working out of the three and sometimes none at all! Things got tricky then, for it was very difficult to judge the distance between probe and drogue.

There were several hazards that came with the flight refuelling – firstly the hose would not trail at all and that was the end of that sortie, although that wasn't dangerous. Next the hose would not wind in at the end of the transfer and that left you with the options of either jettisoning the hose or landing with it trailing. The latter was the most dodgy, as you had to stop the aircraft before you wore right through the hose for the cloud of sparks coming from it would ignite the fuel which would have leaked out. The hose itself held 60 gallons, so it would have been a fair little fire. I had to do this twice – and both times managed to stop in time. It did bring out some spectators to watch the display, though. The trailing hose gradually wore through, creating a fine display of sparks! I also dropped the hose twice, once at Goose Bay in the middle of a snowstorm which was just a little scary because you had to get down near the deck to do it. The other occasion was at Marham and the weather was a good bit better.

Probably the biggest hazard, though, was on the receiving side. If there was any turbulence about, or there was a sudden movement between the two aircraft, the hose would whip, just like whipping the end of a skipping rope, and it would take off the refuelling probe as clean as a whistle. This left the probe stuck in the drogue and it would pour fuel until the operator turned off the fuel in the cockpit. There was so much fuel pouring out under pressure that you could not see a thing out of the front and sometimes it would over-fuel the two inboard engines and cause them to flame out. This was not a good thing. That only happened to me once, but it was enough. On the other occasions I lost the probe, I managed to pull away in time to escape with all the engines still going.

By this time I had accumulated enough points to be offered a 'married quarter' on the base and we were lucky to get a good-size three-bed detached house right beside the fence to the airfield itself. I used to park my aircraft about 100 yards away. It was very convenient, I could walk to the squadron in about ten minutes and there was also lots of company for Eileen and the kids. Marham was also a very 'social' sort of station and we went to and hosted some pretty wild parties, not that we drank, of course! All in all life was pretty good.

The flying was fascinating, there were so many different facets to the refuelling game and much of it took us abroad. It was hard on the families really, but the mutual support from all the other families was great. It was good to know that this was there.

As usual, there are always a couple of incidents that stick in the memory. We had by this time trained up other pilots from other aircraft to flight refuel and one major exercise was to ferry a squadron of Javelins out to Singapore to take part in a big exercise. We worked with two tankers supplying four Javelins on each leg of the trip and we stopped at Cyprus, Aden and Gan before reaching the Far East. Gan was a speck of a coral island on the Equator in the Indian Ocean, and this turned out to be the memorable leg. We left Aden and were about halfway to Gan when one of the Javelin crews lost their oxygen supply, called an emergency and dived for a safe height to breathe. We had a quick discussion and Pete Butler, flying the other Valiant, hurtled down after the Javelin to refuel them while I stayed at height to supply the other three. This meant that we were all going to be short of fuel by Gan, but we calculated that we could do it. When we got there, we sent the Javelins ahead to land first and Peter and I checked our remaining fuel to see who was more desperate to land first. I had the lesser fuel, having kept three aircraft supplied, so in I went. That was when luck played its hand, for when Pete landed just after me, he burst the tyres on the starboard wheels and careered off the runway, starting to sink a bit into the soft coral. The runway was blocked for six hours and goodness knows what I would have done had I not landed first. There would not have been enough room for me to land on the runway and I did not have enough fuel to reach Colombo, our nearest diversion.

As one would imagine, there are lots of tales to tell from 214 Squadron, but there was one exercise which made the national news and TV and certainly sticks in my mind. Before the war one of the bold aviators of the 1930s was Alan Cobham and his dream was to fly to Australia non-stop. He invented an air-to-air refuelling (AAR) system, but no aircraft of the time was good enough for the job, so he formed Flight Refuelling Ltd and worked on developing the system that we were now using. He still harboured the idea of Australia non-stop and I think he persuaded the RAF to try to do it. So, the planning started and the mighty Vulcan was chosen for the glory although the Valiant could well have done it too. Of course it was 617 Squadron who got the plum job, so we trained up the crew and started to work with them seriously. The captain was Sqn Ldr Mike Beavis and the nav was none other than Clive Taylor, who was an old friend of mine from our Canberra days. We plotted the route and it was decided that refuels were needed at Cyprus, Karachi and Singapore. The Karachi one was the trickiest, for it was a night refuel and our flight commander, 'Mac' MacFurze was given the job of lead tanker with me as his back-up (there were two tankers at each location). The route was tested as far as Singapore using one of our own aircraft and all went well. So, all was ready for the big detachment and we went out to Karachi about a week early to set up everything and to get our ground crew settled.

An unusual social event occurred here. The Pakistan Air Force (PAF) invited us to a game of cricket at the weekend and, having just about enough players to make a team, we agreed. The game was to be played at the PAF Academy ground, starting just after lunch. What we did not expect was that the ground was a big place, with proper stands, a big scoreboard and about 3,000 spectators! The match had been billed as the PAF versus the RAF. They were far too good for us and declared at about 220 for 6. I had bowled a lot and suddenly the earth spun round and I spun in! It was very hot and it had got to me. I was taken off the pitch to the pavilion where I was filled up with water and salt tablets and after some twenty minutes went back to rejoin the fray. We didn't disgrace ourselves though, we were not as easy to get out as they thought and battled out

Alan's crew on their way. (*Alan McDonald*)

an honourable draw, scoring 160–odd for 7. Our hosts were very convivial and there was much curry and beer to follow.

The day came for the big event and our job was to meet the Vulcan over the Karachi beacon at around midnight, the final timing depending on what winds they had encountered. We took off in good time and were set up in our orbit waiting to make the initial contact. When we heard from them and got an ETA, Mac gave the word for us to trail our hoses and I took up a position about 400m out to his right. This is when the problems started, no way would our hose trail out. Wally Kerans did everything he knew in the back and I was throwing the aircraft around to try to dislodge the darn thing. Meanwhile, the Vulcan had made contact with the lead tanker and started to take on some fuel. This was when the next snag occurred, for every time the pump was turned on to transfer fuel, it disconnected the Vulcan from the probe. Several attempts were made, but to no avail and it was decided to pass the job over to us. Fortunately, we had just managed to trail our recalcitrant hose and were ready to go and Mike Beavis came over to us to make contact.

There was something amiss with his probe settings, though, and the same thing happened when we turned on our fuel pump, the two aircraft disconnected. Some time had passed by now and we were approaching India from Pakistan. The Indians were not too keen on us overflying and had given permission rather reluctantly, so they sent up a Canberra to escort us and make sure that we got up to no mischief. It was decided that the Vulcan should stay with us and that we would refuel him using gravity feed only. The snag with this was that it took about forty-five minutes to transfer 40,000lb of fuel and it was going to be a much longer trip, taking us much further into India. It was now that our real troubles started. Ahead of us was the biggest line of thunderstorms that you have ever seen – the monsoon! It was a solid wall of cloud, way above our height, and there was constantly flashing lightning. It looked horrifying. The air was full of static electricity, with St Elmo's fire crackling all round our metal window frames, and if you lifted your hand off the controls, a giant spark leapt out to the nearest metal.

There followed a three-way discussion on what we should do. I was unsure that Mike Beavis could keep in contact through the turbulence – and I was not too keen to face the turmoil myself. But there was much prestige riding on this trip and it was decided to have a go at it. We reckoned that we could miss the very big thunder-cells by spotting them on our radar and trying to gently fly around them without throwing the Vulcan off the hose – at least it sounded good in theory. So, in we went. The other tanker, plus the Indian Canberra wisely decided to stay on the western side of the monsoon. We weaved our way gently through the huge storms, it was very rough and I do not know how Mike Beavis kept in contact, but we gradually gave him his full fuel load. By this time we had flown right through the front into clear air so we wished each other 'bon voyage' and off he went to become the first-ever aircraft to fly non-stop to Australia.

I had a different problem, because the Indians would not let us land in their country unless it was a dire emergency, so I had to go through the fireworks zone again. It was a bit easier going back, for I had no Vulcan hanging on to our hose, but it was still pretty horrific. At last we made it through and came out into the still, clear night on the eastern side of the weather and, fortunately, Mac was still there waiting in orbit. That was lucky because I had used a lot of extra fuel and it would have been a bit tight to get back to Karachi, so I took on 10,000lb from him and we settled down to a nice quiet trip home after all the excitement.

We landed about 3.30 am and I think the adrenalin and a few beers were still keeping us going at 5 o'clock! It had been an eventful night. The whole trip was a huge success and the Vulcan crew were treated like heroes – Mike Beavis rightly got an AFC and the rest of the crew got Queen's Commendations. As for 214 Squadron, we hardly got a mention and we were corporately a little miffed. I think that only our squadron commander — one Wg Cdr Beetham — was invited to the big presentation dinner. One person did not forget us, though, and that was Sir Alan Cobham. He came to a big dinner night in the mess and presented us with a big silver globe of the world with the route and the refuelling points engraved on it. We had achieved his lifelong ambition to get to Australia non-stop.

As a postscript to this tale, years later, just before I left the RAF, I sat in the office of Air Marshal Sir Mike Beavis at Brampton and we relived that night for half an hour. It had left a big impression on us both. As another little aside to this, I have just read 'Mac' McFurze's obituary in *The Daily Telegraph* (February 2012). Very unusual for a post-war airman to get that sort of mention.

Life at Marham was still pretty good. We had to contend with the arctic weather of 1963 in our quarter, which looked directly across the airfield to the North Pole. There was no central heating and we kept warm with coal fires, portable and dangerous oil fires and thick sweaters. It was also this winter that I did a sea-survival course when I was thrown into Plymouth Sound in just my flying suit and a lifejacket. I think the sea temperature was just five degrees centigrade! I have never been so cold. After I was dragged out of the dinghy I think I shook uncontrollably for three hours. The next day we were supposed to do it all again in multi-seat dinghies, but the entire course rebelled and declined. Fortunately, sense prevailed and we did the rest of the survival course in a giant pool that the navy used to train their personnel.

Soon after this I got my posting notice to go on to the Vulcan and I was very lucky because just afterwards the snag with the Valiant main spar was discovered and the aircraft were all grounded. It was a strange snag, for when small core samples of the main spar were taken, some were fine and others came out as powder. The Valiant that I had flown on the Australia record run was one of the bad ones – glad the wing didn't fall off in the middle of the monsoon storms! I was lucky to have got my posting for there were a lot of Valiant aircrew hanging around with nowhere to go. It was a very unhappy end to what had been a very good aircraft that did everything you could ask of it. I enjoyed my time on the Valiant, some eight years in all, and must have been one of a very few who flew it in three different roles – PR, straight bomber and, best of all, the very early and exciting days of flight refuelling.

Chapter Seven

214 SQUADRON BEFORE TANKING

Roy Monk joined the RAF in January 1951, age sixteen years, as a Boy Entrant at RAF Cosford. After an intensive eighteen months course he passed out as an engine mechanic. At the ripe old age of seventeen, now an SAC, he was posted on to RAF Upwood. He first worked in the Aircraft Servicing Flight (ASF), followed by the propeller bay, power plant bay and finally building up the Merlin engines that powered all the aircraft that were based on the station. Roy was meticulous at observing, and recording, both his daily activities and what went on

Roy Monk, corporal engine fitter Castle AFB, California.
(*Roy Monk*)

in the hum drum of station life. The evidence comes from a multitude of diaries that he kept during his service in the RAF.

Roy recalls:

I was promoted to corporal very early on and found myself posted internally at Upwood to 214 Squadron. In 1954 I experienced my first overseas detachment. I was sent out with the Lincolns to Eastleigh, Kenya for six months for the Kenyan Emergency. We were told that our job was to bomb the Mau Mau. All our Lincolns had the mid-upper turrets removed but they still retained both the front and rear gun turrets so as well as bombing the Mau Mau we also followed up the attacks by low level strafing runs.

However, despite all our efforts I don't think that it was all that successful. I reckon that we destroyed more trees and animals than anything else! In December we returned to the UK, only to find that our squadron was to be disbanded. I was again posted, this time across to 49 Squadron. The following year I found myself detached once more to Kenya. However, this time the operation was different. Even the Government had decided that the bombing campaign against the Mau Mau had not been effective. Furthermore, it was also considered to be too expensive. Therefore, on 49 Squadron we were tasked only with dropping leaflets asking the terrorists to surrender. In July 1955, the Lincolns were deemed to be no longer required and while one half of my squadron's aircraft and men were returned to Upwood the other half, four Lincolns and men along with me posted, were detached to Aden to bomb, machine-gun and launch rocket attacks against the Radfan in the surrounding mountains. Unfortunately my stay was cut short as I ended up being pushed out of the back of a lorry while getting out, which resulted in injuring my leg. After a spell at RAF Khormaksar it was decided to send me home and so I flew back to Upwood in a Lincoln.

Not long afterwards I suddenly found myself assigned to the latest aircraft in the RAF. It was the Valiant. In preparation I was subsequently sent off to Rolls-Royce at Derby to undergo training on the Valiant engine course and to learn about the Avon turbojet engine. On successful completion of the course I was posted to RAF Marham and ready to join my old 214 Squadron when it reformed.

> 214 Squadron at Marham was officially re-formed on 21st January and before some of the aircraft arrived. Roy arrived on 10th January 1956 eleven days before the squadron's first Valiant. He continues:

I was held for a short time on the ASF, while awaiting the re-formation of the squadron. I noted that one Valiant was already on the station even before the squadron – WZ378. Over the next few weeks, with me having joined the squadron on 13th February, I noted down the serial numbers and dates of each of the other 214 Squadron Valiants as they arrived. They were sporadic and, not necessarily, in a logical order. They were namely: WZ381, WZ379, WP211, WZ377, WZ393, WP212, WZ395, WP223 and WZ397. The complement of squadron aircraft was soon complete and a great deal of work was carried out by us with such jobs as cleaning the hangar, jacking up aircraft and checking the nose wheels, building up trestles, checking tools, endlessly, so it seemed, etching tools with the squadron markings and making up new tool kits. Whilst a lot of the jobs were monotonous they were essential to building up a squadron with new aircraft from scratch. From time to time I also ended up going over to 214 Squadron's dispersal on the other side of the airfield to clean out buildings on the dispersal and to light the coal fires. In between times it seemed to me that I was also constantly sweeping our hangar floors.

I always started off my diary each day with a quick line about the weather. The 8th February 1956 was definitely memorable and still sticks in my mind to this day. The weather was foggy and raining and that day six Hunters crashed while trying to land at Marham. I knew that one pilot was killed. I didn't see any of the crashes myself as it was too foggy but I did see one of the successful Hunters taxi past me on the perimeter track. I also saw one Lincoln, also diverted in, trying to land but it was clearly too difficult and it had to depart for somewhere else.

Under plane crashes in *The Guinness Book of World Records 2014* it states 'Most fighter planes lost in one sortie'. It goes on to say that 'The Royal Air Force lost six of eight Hawker Hunter combat aircraft on 8th February 1956. At RAF West Raynham they were due to undertake combat training at 45,000ft but they were diverted because of poor weather to RAF Marham. A sudden deterioration in visibility precluded visual approaches and there was not enough time to effect spaced radar approaches. Four pilots ejected when they ran out of fuel. One crash landed and survived and another pilot was killed. Two aircraft landed successfully.'

In February I was vetted for security. However, to me a more important day that month was when it was announced that we would all get a pay rise. Every Thursday we attended pay parade and as a single corporal engine fitter I was now being paid the princely sum of £1 16s 0d per week. Nowadays it seems a laughable amount. However, it was enough for me to live on and enjoy my free time. One day I went to Norwich see one of the big bands. It was all the vogue at that time and the highlight was Stan Kenton and his orchestra. Also not being much of a drinker, and keen on films, I used to go to the Astra Cinema on the station regularly. Usually twice and sometimes three times a week. Other off-duty moments, during the week, I spent in the NAAFI or the Corporals' Club, where I was vice-secretary, playing snooker. Although I lived in one of the airmen's blocks during the week I went off to Dartford on every available weekend to see my family and fiancée. We RAF servicemen regularly used to catch the Fenman train from Downham Market to Liverpool Street station, London. I then went on to Dartford.

On 6th April I remember seeing a flypast of ten Valiants. They were practising for the visit of Bulganin and Khrushchev and was certainly an unforgettable sight. A few days later I saw another formation of ten aircraft along with Canberras and Hunters. Three of the Hunters did aerobatics.

Khrushchev inspecting Valiants at Marham.
(*David Sykes*)

Between 18th and 27th April Nicolai Bulganin and Nikita Khrushchev came to Britain. The Queen and Duke of Edinburgh visited Marham on 23rd July 1956 for a review of Bomber Command and to present 207 Squadron with its new standard. On the day of the Russian visit there was a flypast of 16 Canberra B6s, 28 Hunters and 12 Valiants plus aerobatics by four Hunters. On the Queen's Review the flypast consisted of 72 Canberras and 20 Valiants. Clearly what Roy witnessed were practice flypasts for the Russian visit and subsequently the Queen's review. **Roy Monk** again:

My weekly pay was gradually going up by now and in May I was getting £3 18s 0d! On 1st October the main talk of the day at work was that of the dreadful crash of Sir Harry Broadhurst's Vulcan at London Airport. On 20th October, along with others on the squadron, I was told by my flight sergeant to pack up my things. We were all going to Malta. The next day we found ourselves on board a Shackleton heading for RAF Luqa.[7]

My next detachment overseas was in 1957 to RAF Akrotiri, Cyprus. On 2nd October I flew to Halfar, Malta in a Beverley then on to Cyprus arriving on 3rd October. The detachment was an exercise with the American 6th Fleet whereupon 214 Squadron was to simulate attacks on the US Navy. We were accommodated in tents for the month's duration. Quite a large encampment too. The reason for the tents was that parts of the airfield and the accommodation at Akrotiri were still being constructed with loads of corrugated sheets around. The security for our aircraft was just some crudely fixed barbed wire around the Valiants which we all had to take turns in guarding. My most treasured memory of that detachment was my flight in one of our Valiants, XD869, on an air test. I was airborne for one and a half hours. It was extra special for me as it was one of the aircraft that I carried out most of the work on and I considered it to be my 'baby'. One of my colleagues even took a photograph of it taking off, with me on board. I returned on 7th November in a Hastings to Luqa, Malta then on to Marham arriving on 8th November to endure a seven-hour wait for processing through customs!

> This was clearly a NATO exercise and it was a not uncommon occurrence for the RAF and US Navy to play 'war games'. Earlier in February 148 Squadron Valiants had already completed the same type of exercise called Green Epoch and in April 207 Squadron Valiants followed with Exercise Red Pivot. In all instances they carried out simulated attacks against the US 6th Fleet testing the fleet's air defences. Roy continues:

Little did I know at that time that one year later that same Valiant that I flew in would crash just after take-off from Marham killing all on board. It was Flt Lt Watkins and crew. As it was on its way to Nairobi there was a crew chief flying in the sixth seat. It was Chief Tech Bob Sewell, a grand chap and well-liked by all of us ground crew. It was quite a blow.

SAC Combat Competition 1958

Early in July I heard that our Valiants were going to be sent out to the United States to take part in the SAC Bombing and Navigation Competition. I'd never been to the States before and as our sergeant engine fitter and another corporal engine fitter had already been I was one of those lucky enough to be selected to go out with the 214 Squadron Detachment. It wasn't long before we all had to go to RAF Innsworth to be kitted out in gaberdine khaki drill (KD) No 6 uniforms for the USA. Back at Marham I had to endure the inevitable KD kit inspection which was closely followed on the same day by the usual vaccinations for overseas. I, like most in those days, felt awful after the injections and suffered a swollen arm

---

[7]   This was Operation Musketeer recounted in Chapter Four.

all the next day. Before the Valiants went off to the States we were re-fuelling them when the one that I was on was struck by lightning. That was no laughing matter and luckily we were on the opposite side of the aircraft when the strike hit the tailplane, went down the fuselage, then down cables, the generating set that was switched off and then to earth. The RAF policeman guarding the aircraft gave an enormous shout and danced around as the residue of the strike had gone through his hobnail boots. On 4th September I packed for the trip and took my kit to Station Flight ready for a take-off the following morning. Unfortunately, although we did take off the aircraft, a Comet, developed compass trouble on the way to Iceland. We had to turn back and be diverted to Lyneham where I spent the night. Next day the take-off was delayed again due to UHF problems. However, we eventually got airborne and I spent the night at RCAF Goose Bay. On 7th September the aircraft staged through Lincoln before finally arriving at March AFB whereupon, although not a drinker, I celebrated with a few drinks on my first introduction to an American PX! We had a good time there. During our stay, apart from work, our off-duty hours were spent swimming, going to the nearby town of Riverside, visiting bars, going to beer calls and enjoying BBQs. On top of all that we also took trips to San Bernardino and Hollywood. We also went to see the American Grand Pix. All in all the Americans made us very welcome. I was quite impressed that the day before the start of the competition, Air Vice-Marshal 'Bing' Cross, AOC 3 Group, Bomber Command who had arrived earlier in a Victor and was there to watch the proceedings, took time to shake hands with the RAF aircrew and ground crew.

In October 1958 Valiants were invited to take part in the 10th Annual SAC Bombing and Navigation Competition at March AFB, California. It was carried out between 13th-18th October and only the second time that Valiants had taken part. Out of a total of ten crews from Marham, Honington and Wittering selected, five were from Marham. Eight RAF crews were eventually picked to take part and were entered as two wings. It is believed to be the greatest number of RAF crews to have taken part before or, in the ensuing years, after in the SAC Annual Competition. Thirty-nine SAC Wings consisting of 156 crews participated. All wings and individual crews competed for the Fairchild Trophy. However, along with the competition for the major trophy there were also two other competitions running at the same time. One class was the SAC B-47s pitched against the B-36s while the other class was the B-52s against the RAF Valiants. The RAF did extremely well against stiff opposition. One Valiant wing in the B-52 class came 7th overall out of forty-one teams, with another 20th overall. In terms of crew placings three of the RAF came 9th, 12th and 30th out of the 164 crews.

Through 1958 and the period that followed 214 Squadron conducted a number of tanking trials. At the same time it maintained its bombing role. However, once the trials were complete, and in due course, the squadron changed its role entirely to that of tanking. In 1959 the squadron established four unofficial, non-stop, long distance records, three from its base at Marham to Nairobi, Salisbury and Johannesburg and one from Heathrow to Cape Town. Each time Valiants were refuelled in the air. The flight to Nairobi that Roy recorded was a distance of 4,350 miles that was covered by a Valiant in seven hours forty minutes. He takes up the story:

SAC Bombing Competition March 1958. (*Roy Monk*)

On 16th March 1959 XD860 departed for Nairobi, Rhodesia for a long distance record. It returned to Marham on 19th March. During both the outward and return flights it was refuelled by other aircraft from our squadron.

My last overseas detachment in the RAF was on 20th May 1960 when I flew out to Karachi via Malta, El Adem, Khartoum and Aden. Our task was to be positioned at Karachi in Pakistan, along with two Valiants and crews. Their job was to refuel another Valiant, captained by Sqn Ldr Garstin from our squadron, while it was attempting to complete a non-stop record flight from UK to Singapore. The Valiant was to be refuelled twice in the air. The first re-fuelling bracket was undertaken by one of our two Valiants pre-positioned in Cyprus while the second airborne re-fuelling was carried out by one of our two Valiants that were pre-positioned in Pakistan. My worst recollection of the visit was the filthy hotel that we were holed up in. There was a vulture sitting on the windowsill inside my room all night. I thought that it was going to carry me off! Not long after the flight, it was in the newspapers, we were told that the time taken for the Valiant to fly from the UK to Singapore was fifteen hours thirty-five minutes. After that record was made a few days later the Garstin crew flew on to RAAF Butterworth, Malaya and broke another record on the return flight from Malaya to Marham. The rest of the year at Marham, and my final year in the RAF, it was business as usual on the squadron. On 1st January 1961 I left the squadron and the RAF to become a civilian.

Chapter Eight

OPERATION GRAPPLE

Operation Grapple began in 1957. The testing of a large megaton-yield weapon had dictated that a special site be found and Christmas Island, a remote coral atoll 2° north of the Equator, was chosen. The tri-service and civilian task force for Grapple was commanded initially by Air Vice-Marshal W E Oulton and then by Air Vice-Marshal, later Marshal of the RAF Sir John Grandy. The scientific director was Mr, later Sir William Cook. Each service had its own task group; the RAF 160 Wing was commanded by Air Cdre C T Weir.

Although the island was inhabited with an economy based on the export of coconut products, it had largely been neglected since World War II. For Grapple, preparations had been started the previous year to construct the support facilities; a 7,000 feet runway, twenty-five miles of roads, a control tower, buildings for weapon assembly and a sea water distillation plant were just some of the building works needed.

The domestic accommodation for personnel, not to mention the indigenous gerboa rats, was all tented but more substantial buildings were provided for recreation purposes. Most of these works were carried out by the army task group of Royal Engineers which included a detachment of Fijian troops. To provide some relief from the somewhat primitive living conditions, the RAF contingent was entertained regularly on the light aircraft carrier HMS *Warrior*.

49 Squadron received eight Valiant B(K)1 aircraft; underwing tanks were not fitted for the first set of Grapple tests. The first (XD818), resplendent in 'all white' anti-flash finish, arrived at Wittering in November 1956. In addition to the heat-reflecting finish, metal anti-flash screens were fitted to windows serving the flight deck, crew and bomb-aiming

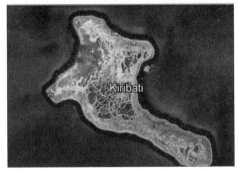

Christmas Island with airfield on north shore.

positions. Cameras were installed in a new tail cone and on mountings within the bomb bay, and the control surfaces were reinforced. A lot of automatic strain and pressure detecting devices were fitted to the airframe and many vents and drains baffled or sealed off. The V Force standard radar navigational bombing system was replaced by a special scientific panel.

The bombs to be tested during Operation Grapple were trial warheads for a proposed range of British thermonuclear weapons: Short Granite, Orange Herald and Purple Granite.

For carriage on the aircraft, these trial assemblies were fitted within Blue Danube casing. The weapons would be released from 45,000 feet and set to explode at 8,000 feet, thus avoiding the fireball touching the surface and causing excessive radiation.

The dropping points for Grapple were planned off Malden Island south of Christmas Island. Before each drop scientists had to make the final checks on the apparatus sited on Malden, which was designed to measure the air blast, heat and radiation levels. They then withdrew and embarked on HMS *Narvik*, *Warrior* and *Messina*.

The hours prior to the release of the Grapple bombs were full of action since before dawn the Shackletons of 204 and 206 Squadrons had to take off to search the danger area to be sure it was free from shipping. In addition Canberras of 76 and 100 Squadrons had to depart Christmas Island on weather reconnaissance and sampling sorties. Then at first light, Hastings of 24 Squadron and Dakotas of 1325 Flight would leave for the target area laden with observers. When 'all clear' messages were received from the task force commander on HMS *Narvik* and the operations room on Christmas Island, the drops could go ahead.

In the case of Grapple 1 it was XD818 that carried the first bomb piloted by Wg Cdr K G Hubbard. His commentary was clear and unhurried as the aircraft made its run-in at 45,000 feet, at a speed of 0.76M. 'Bomb gone' came the brief announcement from bomb aimer Flt Lt Eric Washbrook, as the Short Granite round fell away. Immediately the aircraft swung in a 60° banked turn to port; this escape manoeuvre would take the aircraft to a position some six miles distant in the thirty seconds before detonation. A stark, blinding flash at 8,000 feet in the Pacific sky signalled Britain's emergence as a top-ranking nuclear power. It was 10.36hrs local time, on Wednesday 15th May 1957, almost six years to the day since the first flight of the Valiant prototype.

The aircraft turned away from the explosion and the crew did not look back until five minutes later, and saw what Hubbard described as 'towering above us, was a huge mushroom-shaped cloud, with the stem a cauldron mass of orange as the fire ball had developed, and the hot gases risen into the atmosphere, progressively fanning out forming a foaming white canopy which can only be compared with the top of a mushroom. This top must have reached an altitude of approximately 60,000ft with ice caps forming.'

In fact the bomb had a disappointingly small yield of 300 kilotons though still twenty times the power of the atomic bombs that destroyed Hiroshima and Nagasaki in 1945. The problem was that the scientists were looking for thermonuclear explosions in megatons rather than kilotons.

Second and third drops of weapons in the megaton yield range took place at the same location on 31st May and 19th June, the aircraft being captained by Sqn Ldr David Roberts and Sqn Ldr Arthur Steele, XD822 and XD823. 49 Squadron returned to the UK in June. In fact the second bomb had a yield of 720kt but failed to meet the expected 1 megaton; furthermore the bomb was actually a fission bomb, though probably the largest ever made. The third bomb was similar to the first one but in fact only had a yield of 150kt. Seven AFCs were awarded to squadron personnel in the 1957 Queen's Birthday Honours.

Grapple 1 explosion.   Celebrations after Grapple 1. (*Clive Cox*)

The *Daily Express* on the day following shot three made it clear that this was the final full explosion and was probably the biggest aerial bomb ever exploded at that time. However there was a suggestion that not everybody in the UK and Australia was in favour of these tests from a cartoon published on that day by *The Daily Telegraph* (see below).

We are lucky to have three accounts of men living and operating from Christmas Island. The first one from the wireless mechanic of XD822, **David Kent**:

I joined the Royal Air Force in March 1956, signing a three-year contract. This was not uncommon in those days. For many, it was a device to avoid the lottery of National Service where you could end up anywhere doing anything and probably in unpleasant surroundings. I had loved aeroplanes since being a small child so the choice of the air force was not difficult for me. During basic training various carrots were dangled in front of you. The best trade training only would only be open to you if you signed a longer contract and

by the time I arrived at No 2 Radio School, Yatesbury, Wiltshire, my term of engagement was now five years. I became one of the first air wireless mechanics to undertake a specialist V bomber wireless course and was posted to 49 Squadron at RAF Wittering in October of 1956. 49 Squadron had been reformed in the summer of 1956 and its first aircraft and crews had been from 1321 Flight. 1321 had been under the command of Squadron Leader Dave Roberts and it had been expected that he would be the new officer commanding 49 Squadron but that was not to be, and Ken Hubbard was appointed.

Twenty-five years later, at a Megaton Club[8] reunion at the George Hotel Stamford, in a conversation with Dave Roberts, he revealed how bitter his disappointment had been at not being offered the command of the squadron. It was clear that he never understood why he was overlooked and that the decision still rankled all these years later; so Ken Hubbard must have had a bigger job to do than we all understood in persuading the crews that he was the right man for the job. The consolation prize for Dave Roberts was that he was made flight commander of A Flight.

After the hectic preparation of our four Grapple aircraft, the ground crews left for Christmas Island in February 1957. The Valiants left the UK at intervals during the beginning of March. For a nineteen year old who had never been overseas in his life, the journey to Christmas Island was an exhilarating, and unbelievable experience. The flights were with civilian airlines, BOAC, from London to New York, United Airlines New York to San Francisco and then on to Honolulu. On the outward trip from Heathrow in a Boeing Stratocruiser I woke at dawn to join in with a discussion about the positioning of the sun in relation to the aircraft – should it have been on the port side or the starboard side? The debate was quickly resolved by the captain of the Stratocruiser who reported to us that the whole eastern seaboard of the United States was fog bound and instead of flying west we were sedately returning to the UK shores and were due to land at Prestwick.

After a night at the Marine Hotel, Troon, we set off once more, this time via Keflavik in Iceland and on arrival in New York, the booked scheduled flights had long disappeared and we were all divided up among various airlines and took varying routes. Ted Firth and I, who were crew colleagues, ended up on a United Airlines flight to San Francisco via Chicago where some delightful flight stewardesses introduced us to the sophistication of Manhattan cocktails – all free of course. We were eventually all gathered up in the Drake Wilshire hotel just off Market Street in San Francisco and after a day and night 'doing the town' which, in my case, involved being driven across the Golden Gate by an ex-pat Brit, we flew to Honolulu the following day.

We were billeted in the US Air Force Base at Hickam and were fascinated when it was explained to us that the bullet holes in our accommodation blocks were the aftermath of the Japanese attack on Pearl Harbor in 1941. We spent a wonderful two days enjoying the delights of the Hawaiian paradise. It was

---

[8]   49 Squadron members who were at Christmas Island.

then south 1,300 miles across the Pacific to another island paradise – allegedly – Christmas Island. Six to a tent, open showers, earth closet toilets, pretty average food, DDT sprayed from an Auster aircraft three times a week to keep the flies down and football results from the UK at 5.30 am on a Saturday morning. Our initial reaction to the heat was one of total disbelief. How could we live in these temperatures never mind work in them. The heat problem was made worse by the total absence of any sun protection. Not for us SPF50; so fair-skinned folk like me suffered. However, by the time the Valiants arrived in mid-March, we were well acclimatised and up for the task.

The first test took place on 15th May 1957 when XD818 captained by Ken Hubbard dropped the first of a series of nuclear weapons.

Orange Herald

31st May 1957 was 'D Day' for the second of the Grapple nuclear weapons tests – codenamed Orange Herald. The aircraft tasked with the mission was XD822 skippered by Squadron Leader Dave Roberts. I was wireless mechanic on XD822; it had originally been delivered to the squadron from Vickers on 9th November 1956 and I had been assigned to its ground crew which was under the leadership of Chief Technician Sam Small. XD822 had undertaken its first operational flight as a 49 Squadron aircraft on 15th November 1956 so the preparations for the day had taken only six months – a not insignificant achievement.

The weather on the morning of 31st May was dry and clear. I was woken from my sleep at around 5.00 am and we made the two-mile trip to the airfield. We had been making the journey four times a day since our arrival on Christmas Island in February but this day was different. It was the first time that I had undertaken the journey in the dark. The roads on Christmas were graded coral and the lights picked out the palm trees and the scuttling land crabs. As we approached the airfield another tropical dawn was breaking and we could see on the squadron dispersal that Valiant XD823, which was to fly with XD822 as the observing aircraft, was already lit up and pre-flight preparations were underway. Our destination was not our usual jet pan dispersal but the Atomic Weapons Research Establishment (AWRE), Aldermaston, compound which was some half a mile away from our dispersal, near the end of the runway.

The aircraft had been taken there two days previously to be prepared by the scientists and technicians for its date with history. The Valiant was surrounded by high screens and illuminated with orange lights. As we entered the security compound, the sight that greeted us was a familiar one. Our aircraft stood illuminated by the orange lights. Bomb doors open and the black and white casing of the 10,000lb weapon clearly visible. We had all seen this many times before. The casing of the practice weapons was identical to that of the real thing and this was the real thing. I don't know about anybody else but I had a feeling of considerable awe and not a little respect, tinged with a frisson of fear.

We all set about our pre-flight tasks. As a wireless mechanic I needed to check the VHF systems. There were three VHF sets in a Grapple-modified Valiant, two for normal voice communications and one which was for telemetry purposes and was specially fitted to assist with signal generation related to the exploding of the weapon at a given altitude. I then needed to check the HF radio. This was a new piece of kit which had been developed by Standard Telephones and Cables. It was called the STR18 and was the first miniaturised crystal control HF transmitter receiver. I then checked the ILS and the radio compass. My final task was to communicate with the tower and ensure that the pilot's and co-pilot's VHF controllers were working correctly.

Bomber Command in those days operated a crew system for ground engineering. Standard ground crew for a Valiant was made up of fitters and mechanics specialising in air frames, engines, electrics, instruments, radar, wireless and an armourer, all under the leadership of a chief technician. The crew system which was disbanded some years later on the grounds of cost played a significant role in the exceptional morale and high standards of team work which existed within 49 Squadron. In our particular case on XD822, from memory, the same crew were together for around eighteen months. When our tasks were completed, we waited for the aircrew to arrive for them to undertake their pre-flight tasks. I confirmed to my crew chief and my AEO, Flt Lt Jock Beattie, that all was in order. All that was left to do was to sign the form F700, the aircraft's engineering log, and leave the aircrew for their date with destiny. 822 taxied slowly out from the AWRE compound onto the runway.

I can do no better than quote from our squadron commander Ken Hubbard's book: 'Valiant XD822 moved majestically onto the runway and commenced its take-off run, becoming airborne at 09.07 hours local time. With undercarriage retracted, she quickly settled into a climb, heading towards Malden Island.' The second live thermonuclear weapon Orange Herald was dropped at 10.44 hours local time. The aircraft and crew returned safely; however this was only made possible by the exceptional skill of the captain, Dave Roberts, who had a major instrument failure during the sharp turn away from the drop zone. It was, in fact, the only major technical fault on the Valiants in the whole of the Grapple trials.

Fast forward to 1994. My wife and I did a trip to Christmas Island. Spent a week touring familiar sights. The rusting relics of Grapple were everywhere. The most pristine relic was the weapons assembly shed which today serves as a store for an islander called John Bryden who supplies fuel and vehicles in support of the island's economy. The deserted jet pan brought back many memories of an unforgettable experience for me.

Publicity these days seems to centre only on the issue of nuclear radiation and its affects, if any, on a small minority. Perhaps people should also remember that it was a fantastic technical achievement for all concerned, not just 49 Squadron but for the Royal Air Force, Army and Royal Navy personnel too.

An enduring legacy has been our Megaton Club and friendships forged over 50 years ago which are still as strong today.

Reference the Orange Herald drop described by David, I was fortunate enough to speak to Alan Pringle who was the co-pilot on the second Grapple drop flight. The aircraft was fitted with a large face remote-sensing accelerometer instead of the normal small integral accelerometers which were usually used for flight testing. The maximum amount of acceleration permitted during the escape manoeuvre turn was 2g to prevent damaging the aircraft. During this steep turn the drill was for the second pilot to call out the

Wg Cdr Hubbard relaxing outside his tent. (*Clive Cox*)

accelerometer readings continuously as the captain carried out the turn applying but not exceeding 2g. However on this occasion one of the wires fell off the accelerometer so Alan had no g to call out. Dave Roberts, the captain, pulled the aircraft into the turn increasing g expecting to hear Alan calling out the g's but there were no calls and the aircraft entered a stall, presumably because of excessive acceleration being applied; luckily Roberts was able to recover from the stall without going into a spin. Just as well he wasn't in a Victor!

The second Grapple account is from **Don Briggs** who was a flight engineer flying Lancasters during the war on a Pathfinder squadron. He was flight engineer at Empire Test Pilots School (ETPS) 1948/50 and was then posted to Manby where he met a wing commander that he had known on his squadron; as a result he managed to get trained as a pilot in 1951. He got posted to 49 Squadron and took part in Suez and Grapple; he was in Sqn Ldr Arthur Steele's crew as co-pilot on XV823 which dropped Grapple 3. Luckily he dictated his Grapple experiences and also his Valiant flying after Grapple which I have left in.

Our squadron commanding officer was Wing Commander Ken Hubbard and he immediately appointed Squadron Leader Arthur Steele as the 49 Squadron training officer whose job it was to work closely with the scientists at the AWRE. We carried out lots of crew training particularly practice bombing; we dropped a 10,000lb Blue Danube dummy bomb and later a similar bomb containing telemetric equipment which could be monitored by the scientists on the bombing range as the bomb was falling. Later in 1956 we flew to Malta to take part in the Suez campaign. A crew from 138 Squadron had gone sick and our crew took their place to carry out a bombing attack on an Egyptian airfield which was Cairo West. As a matter of fact I used to stage through this airfield during my time on Yorks on Transport Command. We reportedly damaged some of President Nasser's MIG 15 aircraft on the ground. This was at night of course and we bombed from 45,000 feet so nobody could get near us at that particular height. Arriving back overhead at Malta we had a near miss with another Valiant which was approaching head on but he was at the wrong height. When it was discovered it was a mistake sending us out to Malta in the first place, Arthur Steele organised a flight back to RAF Lyneham aboard a Comet 2 of Transport Command.

We continued our Grapple training and were re-equipped with our brand new Valiants which were to carry out the live H bomb drops. We were flown down to the Vickers factory at Wisley and proudly collected our own Valiant, XD823, to fly it to Wittering. However we were not pleased at the prospect of being separated from our families at Wittering and to spend an unknown number of months on a Pacific island living in tents.

On 11th March 1957 we flew our Valiant to RAF Aldergrove to top up our fuel tanks and then took off again to fly the Atlantic to Goose Bay, Newfoundland. This was the first time I had crossed the Atlantic, and it was a five-hour uneventful flight. Although the temperature was around zero most of the snow had gone. I seem to recall that the Royal Canadian Air Force people gave us a warm welcome. The following day we flew another five hours due west and night-stopped at an RCAF Base called Namao near Edmonton, Alberta; this was a Canadian Air Force station, some civil airline traffic used it as well. The following day we flew to Travis Air Force Base near San Francisco; a relatively short flight, just over three hours and we flew parallel to the Rocky Mountains which was a marvellous sight. I knew the commanding officer of the RAF detachment who was my ex squadron commander on Canberras. He had a ground handling party there; he very kindly drove us all out to a very good restaurant where I enjoyed my first meal on American soil.

The next day transport was arranged for a sightseeing trip to San Francisco and the following day we planned our flight from Travis AFB to fly the Pacific Ocean leg to Honolulu. Fuel usage was critical as head winds were normally encountered on that westbound leg. We found that without having underwing tanks fitted we could not make it with the average maximum head winds exceeding 60 knots[9]. Fortunately, the headwinds were not that strong so we made Honolulu International Airport with a good fuel reserve. It was my turn to land the aircraft and the approach was made to the runway over Pearl Harbor. Quite a sight! It was St. Patrick's Day when we arrived and the officers club at the air force base had a big function, they always celebrate St. Patrick's Day and our crew were all invited. The following day we visited the famous Waikiki beach with an excellent view of Diamond Head and it only remained now to complete the flight to Christmas Island by flying the 1,000 miles leg due south of Honolulu.

Our ground crew were already there to see us in and after the long flight from the UK there was very little that had gone wrong with our aircraft, just

---

[9]   There were no underwing tanks for the first Grapple aircraft.

a very few minor defects. The 7,000-foot runway had been laid by the Royal Engineers and was completed ahead of schedule. The runway surface was quite good. Living in a tent, there were two officers to quite a large ridge tent, it was a bit rough and we slept on Safari canvas camp beds. The officers' mess was a marquee and the food was excellent. There was a Hastings aircraft flying fresh food from Honolulu, which was supplied by the Americans about two or three times a week so at least we were well fed. We quickly settled in with a fair amount of flying, mainly practice flying at Malden Island, a small uninhabited atoll south of Christmas Island and where the live H weapons were to be dropped. We also dropped 10,000lb inert bombs at Malden and the scientists on board HMS *Narvik* had their telemetry equipment to monitor the signals transmitted by the bomb as it fell, but being inert there was no explosion.

We had plenty of spare time and almost every day we swam in the crystal clear water just inside the coral reef. We were all keen on snorkelling and the NAAFI shop stocked all the equipment that we needed. We were not allowed to go on the reef or were advised not to when the tide was in; the tidal variation was quite slight in that part of the Pacific — only about two or three feet — but nevertheless at high tide the currents were so strong on the reef that I believe one could be swept out to sea. There were big sharks out there we were told but in fact there was a helicopter search and rescue service always on standby in case anybody did get washed out to sea. The domestic site was about four miles from the airfield and we were bussed to work each day and back for lunch etc. The roads were graded and rolled but the road sank gradually and after a very heavy shower big ruts formed and the graders had to come out and do it all again.

On the island there were millions of land crabs and they used to come out at night in their thousands and many were run over by vehicles but the next morning there were no signs of any; they had all been eaten by their brother land crabs. The land crabs were a nuisance and used to crawl into our tents at night; if one crawled under the safari bed there was only a few inches ground clearance if you were lying in bed and it gave one a bit of a fright if a land crab crawled underneath but we had access to loads of driftwood that washed up on the beaches and we used to get the station carpenter to not only make walls to go at each end of the tent to keep the unwelcome visitors out, but he also used to make a base to fix the bed to so we were higher off the floor. While we were out there of course we did all our own laundry and ironing and mainly wore casual Hawaiian shirts when off duty. We sent pictures home of our crew relaxing on the white sandy coral beach; it must have looked like a South Sea island paradise to the people at home and you could see the coconut palms and shallow lagoons but to us it was 16,000 miles from home and we all missed our families.

Our detachment included two search and rescue Westland helicopters as I mentioned earlier. One of their duties was to deliver food rations across to the other side of the island to what was called the Port of London where there was a small military army and air force detachment. The helicopter boys used to invite the odd one or two of us on a sightseeing run with them on these ration runs. It was very enjoyable flying at quite a low height probably 100ft or less above the shallow lagoons and now and again we would see a manta ray or a basking shark.

On 15th May 1957 our CO Wing Commander Ken Hubbard and crew dropped Britain's first H bomb at Malden Island and this was followed by a second nuclear bomb dropped by Squadron Leader David Roberts and crew and we did the grandstand for his drop which meant flying in loose formation behind and below the bombing aircraft. When he opened his bomb doors we carried out a steep turn to minimise the shock wave of the detonation. It was not certain that we would be required to carry out a live drop that was No. 3 in the series. We were granted a week's leave while politicians and scientists made up their minds. Every week a Hastings aircraft flew to Australia calling at Brisbane. I managed to cable my brother Norman who lived in Brisbane and let him know I was arriving at a certain time. The rest of the crew actually went up to Honolulu for a week and they stayed at an American base quite close to Waikiki beach. My flight on the Hastings was approved and I flew as supernumerary crew. Our first stop over was Nandi airport on the island of Fiji and the hotel at the airport was excellent.

Since leaving Christmas Island we had crossed the International Date Line and gained a whole day. So we had two Saturday nights. The next day we flew from Fiji to the Royal Australian base at Amberley, Brisbane. Brother Norman was there to meet me and brought some of Brisbane press as well. Next day our pictures appeared in the newspapers around Brisbane. Brother Malcolm who was also in the RAF stationed on a Canberra detachment at Melbourne managed to get some leave and he flew up to Brisbane airport so we had three brothers together. After three very enjoyable days with Norman and Malcolm I joined the Hastings crew again and we had a good flight back to Christmas Island landing at an airstrip on the island of Samoa, a beautiful south sea island. The native children entertained us by the side of the aircraft with song and dance, very nice, it was really beautifully done. It was only a short stop before taking off again. We resumed flying in the Valiant and carried out inert drops again at Malden Island and with lots of air tests on our aircraft XD823.

Then came the big day for our live drop – 19th June 1957. The aircraft was checked and double checked and surrounded by screens so no one could see what was going on. The Aldermaston scientists made the final checks of circuits and when they were happy the bomb doors were closed. So we took off and Arthur Steele climbed the aircraft to 45,000ft as we headed south to the Malden dropping zone. We carried out a dummy run with all the blackout screens in the cockpit in place and the rear crew compartment also blacked out. Arthur Steele flew the aircraft for a live run. Flight Lieutenant Wilf Jenkins was our bomb aimer and the live H bomb called Purple Granite was released. The captain immediately executed the escape manoeuvre to the left pulling the exact amount of g force through 135° and then rolled out and held that heading steady while the shock wave caught up with us. In fact there were two shock waves; one was a direct air blast and the other was the shock wave that bounced off the surface of the sea. This was a considerable shock, the aircraft shook quite badly. Rearward-facing cameras in the tail cone recorded pictures of the detonation. We then took down the screens to see the enormous mushroom cloud that reached from 8,000ft to the stratosphere in a very few seconds, an awesome sight. After landing back on Christmas Island Arthur Steele was given an immediate bar to his Air Force Cross. I think our CO said to the rest of us, well done chaps. The powers that be decided that sufficient information and test results, giving the yields of the three H weapons, had been achieved. The government declared that Britain was now a nuclear power and there was no requirement for further tests for a while. That meant that air and ground crews of 49 Squadron could return to the UK. Our beloved XD823 was air-tested several times before the long flight back to the UK.

Five days after dropping our live weapon we took off to start the journey back to the good old UK. When we stepped into the officers' mess at RCAF Goose Bay what should we see but British newspapers with our H bomb drop accompanied by a wedding photo of Edith and myself, supplied I think by my mother. We were very glad to be back at our base at Wittering with our wives and children there on the dispersal to greet us.

Having settled back at Wittering and back to duty after a lovely leave with my family at my parents' home in Yorkshire, I flew my last sortie with Arthur Steele on 20th September 1957 and was replaced by another co-pilot. Arthur and crew flew out to Christmas Island again, Arthur taking charge for the second series of tests known as Grapple X. I had been selected for a Valiant captains' course and successfully passed the interview with the Air Officer Commanding 3 Group Bomber Command, Air Vice-Marshal Cross, who authorised my captains' course on Valiants. Whilst waiting for my course at RAF Gaydon I flew with several different captains but settled down with Squadron Leader Ulf Burberry and crew. We flew to RAF Luqa in Malta for a 'Sunspot' detachment lasting about three weeks. Strangely enough although I did not enjoy my visits to Malta we did more sightseeing on this one and I found it to be a more pleasant detachment. Ulf Burberry was a great bloke to fly with, he bent the rules now and again and allowed me to fly from the left-hand seat several times. This was not normally allowed until the co-pilot had gone through the course as captain.

At Wittering I was appointed officer in charge of station flight. I had two Anson aircraft each of which could carry six passengers, and a Chipmunk to look after. In between Valiant trips I flew lots of trips in the Ansons and kept my hand in doing aerobatics in the Chipmunk. In March 1958 I started my captains' course on the Valiant at RAF Gaydon. I knew the aircraft well and after flying in the left-hand seat with an instructor in the right I flew in command with my own crew after nine hours instruction. It made me feel proud to fly this four-engined V bomber as pilot in command. After flying it at night and passing my instrument rating test we completed the course and were posted to 138 Squadron at RAF Wittering. This meant that Edith and the children could stay in our married quarter at Wittering. My CO in 138 Squadron was the famous Pathfinder Wing Commander Tubby Baker!

My crew and I flew some very interesting trips, including one to Salisbury, Rhodesia as it was then called. We overnight stopped at Nairobi and on that occasion we landed our Valiant on a mud runway at Eastleigh which had been graded and rolled. It wasn't a bad surface, just rather dusty. Although the runway was 8,000 feet long we had very little to spare when taking off the next morning. This was because Nairobi is 5,300 ft above mean sea level where you get reduced thrust from the jet engines. On the return trip we landed at Nairobi airport (Embakazi) for our return home and a night stop there before flying the Sahara Desert leg to Idris airport, Tripoli, named after King Idris. These overseas flights were an important exercise in showing that we could operate as a crew without the support of ground staff, although we did carry a sixth crew member, that was our crew chief who knew the technical side of the aircraft very well. Moreover, and importantly, we were a nuclear bomber crew as part of the deterrent during the Cold War period. It was my responsibility as captain to make sure we completed the quarterly training requirements and we had already attended a course at the Bomber Command Armaments School (BCAS) on the type of atomic weapon that we would carry in the event of an alert from a Soviet attack. We would be operating from dispersal airfields around the UK. Each V bomber crew had to put in a certain number of hours of target study and we had to memorise three separate targets in Russia but that was for the quarterly period. The navigation bag contained all the target material and was drawn from the vault by two crew members. We all had a high level of security clearance and could only draw the target documents on a two-man basis.

The Valiant was a great pleasure to fly and I was extremely proud and privileged to take on the responsibility for the safety of the crew and the aircraft. In February 1959 I was given the opportunity to take my crew on what's called a western ranger to Canada and then on to America. Goose Bay, Newfoundland was in the grip of winter and we landed on a snow-covered runway. Taxying the aircraft was very tricky as the nose-wheel steering was quite ineffective on the ice, it would simply slide sideways. When it came to departure the following day they wouldn't let us go because the runway at Offutt AFB at Omaha had six feet of snow on it so we stayed at Goose Bay in the meantime and literally sat on the ground for a few days. We were able to keep our Valiant in a heated hanger and spent three days waiting for the Offutt runway to be snow cleared.

We then flew from Offutt to Goose Bay and after a night stop tackled the final leg to the UK. The remainder of my tour on 138 Squadron consisted of lots of navigation which included astro navigation and blind bombing using the NBS. Finally we took part in a one-month detachment to the Royal Australian Air Force base at Butterworth, Malaysia near the island of Penang. The flight back to Wittering involved landings at Ceylon, Karachi and Cyprus. I didn't know at the time, but I was going to be stationed in Cyprus as a Vulcan squadron captain later in my air force career. And so ended my tours on the Valiant as co-pilot and captain which lasted almost six years. My next posting was to be on the Operation Conversion Unit (OCU) as a member of staff at RAF Gaydon in Warwickshire.

To return to Operation Grapple, clearly more development was required of the UK thermonuclear weapons and a further series of three tests were carried out with 49 Squadron Valiants, code named Grapple X, Y and Z. Because a solution was urgent and as a result of the earlier tests it was decided that with the same explosion height of 8,000ft it would be safe to use a target on Christmas Island itself with an aiming point on the south-east tip of the island only 20 miles from the airfield.

Grapple X took place on 8th November 1957 and was spectacularly successful; it was a two-stage device, a trigger of fission followed by a fusion explosion; the yield was 1.8 megatons but used a lot of expensive uranium isotope as the trigger; Sqn Ldr Barney Millett and crew in XD824 carried out the test.

Bill Evans, who kept such a splendid diary for Operation Buffalo, has also related his memories of Grapple. His account covers the early tests but includes Grapple X and Z.

Operation Grapple was a very much larger undertaking than Buffalo. The original intention by the UK Government and scientists was for a one-off series of tests and therefore the facilities at Christmas Island were originally planned and kept to a minimum. However, Grapple developed into a whole series of tests beginning with Grapple, then Grapple X, Grapple Y and finally Grapple Z. Of this four-test series I attended three of them, the initial Grapple and then X and Z, missing out the Y test. These tests were carried out over a period of two years from the beginning of 1957 to the end of 1958. Owing to the length of them I did not keep as concise a diary as I did for Buffalo. I have souvenirs of the events and travels and a few notes so my recollections are from memory but still quite vivid.

On Tuesday 25th February 1957 we departed London Airport by BOAC Stratocruiser for New York Idlewild via Shannon Airport, Ireland for a fuel top-up. The initial flight was fine until we reached American shores where we were unable to land due to fog and had to return to London (see also page 78). Arriving back in London there was an immediate change of aircraft and off again to New York, hours and hours of flying, very boring. From New York after a day and half stay it was then on to San Francisco, via Chicago by American internal airlines. At San Francisco we were accommodated for two nights at an American naval base on the outskirts of the city and visited some of the 'dives' on the Barbary Coast. Next stop Honolulu and then on to Christmas Island arriving 4th March.

Because we were mixed with civilian passengers on the way out we wore our civilian clothes. Our jackets, ties and trousers after a week on the island developed a white mould practically all over and the cloth rotted. We were reimbursed by the

MOD on our return to the UK. Subsequent trips were in uniform which had been thoroughly dry cleaned before departure.

First sight of Christmas Island revealed palm trees and brilliant white coral in a deep blue ocean. Sun glasses were an absolute 'must' against the dazzle from the coral. Atmosphere very hot and humid but quite a strong breeze. The accommodation and facilities were rather basic. We were 'housed' in rather ancient tents that had originally been in the Canal Zone of Egypt and were not entirely waterproof. Coconut matting covered the floor, beds were the low canvas camp type, and a small single wooden locker was provided for our clothes; that was it. Later we constructed wooden rafts approximately 2ft high to fix the beds to and give some shelving space underneath. Land crabs scuttled about and boards were soon placed at the tent entrances to stop them invading inside. Flies were everywhere. During the whole of the Grapple series and possibly afterward, an Auster aircraft sprayed fly killer up and down the camp and airfield area from daylight to dusk every day except Sundays.

What a wonderful job that pilot did. However, the fly population never appeared to diminish! Food was consistently awful and very short rations – we would volunteer for extra shift work in the evenings just to get an extra meal – no food of any sort was available from the NAAFI and unfortunately the land crabs were inedible and Red Cross parcels did not materialise. Toilets and showers were open-sided corrugated sheds. Showers were warmed sea water, therefore special soap was required as normal soap does not lather in sea water. Fresh water was at a premium.

The first Grapple operation was over a six-month period for 49 Squadron; Grapple X ,Y and Z were slightly shorter periods. I lost just over a stone in weight on my first visit to the island, from 10st 10lb to 9st 3lb. The shortage of food applied to everyone, scientists, army, Royal Air Force – everyone. The food shortage and poor accommodation was so dire that the civilian scientists refused to take part in the tests after Grapple X unless conditions were drastically improved, which they rapidly did.[10]

An intense period of work was carried out to prepare the Valiant aircraft for the H bomb tests. Four aircraft were converted, ailerons, elevators, and rudders were removed to fit anti-flash seals, i.e. black rubber to white silicon. Modified tail cones were fitted to house cameras to record the explosions, all dark surfaces changed to lighter colour materials – many telemetry and extra stress gauges and standings and not in hangars so a very strict eye had to be kept on the weather which could change from brilliant sunshine to a ferocious tropical storm within minutes. At the end of all the preparations and three days before an aircraft was due to carry out a 'drop', all the squadron personnel set-to and cleaned the aircraft top to bottom, nose to tail with Daz washing powder to erase all exterior dirt and then equipment were fitted. All this work was carried out in the open and it was hard graft.

---

[10] *Britain and the 'H' Bomb* by Lorna Arnold, page 169 – 'Island Hardships'. She was a public servant and became the historian of Britain's nuclear project.

D Day 15th May 1957 arrived and the first H bomb test at Malden Island, four hundred miles from Christmas Island was achieved.

Prior to the aircraft getting airborne from the island with the weapon on board, I was detailed as a corporal airframe fitter with an intimate knowledge of the Valiant aircraft airframe to be on standby with a 'boffin' sitting in a helicopter with its engine running. Strapped to my back were two large compressed air bottles with a pipe leading to a 12-inch circular saw. This equipment was to be used to cut into the Valiant bomb bay in the event of it crashing on take-off on the island and to allow the 'boffin' access to the weapon to do whatever he needed to do. I was twice detailed for this rather unenviable duty, wearing only a coverall boiler-type suit, bush hat and shoes. No hard hat or mask or gloves. Thankfully there never was a crash, not in the whole of the Grapple series. In later take-offs the fire and rescue services did this job, much to my relief!

Three tests were carried out during Grapple at Malden Island and then when all the aircraft were de-modified we returned to RAF Wittering. Instructions for Grapple X arrived at the Squadron in September 1957 because the Grapple tests had not produced an actual H bomb. However, due to shortage of time and cash and because of the looming moratorium on testing, a decision was taken to carry out the future tests off the end of Christmas Island approximately twenty nautical miles from the airfield. I went on the Grapple X detachment travelling this time from London to Keflavik [Iceland] and then over the North Pole to Christmas Island. We were conveyed by Tiger Airlines – four-engined Super Constellations. On this trip we were issued with small certificates by the airline as members of The Arctic Circle Club. At this time Asian Flu had reared its ugly head in the UK and unfortunately two of our personnel had the 'bug' when we left London – twenty-four hours later landing at Christmas Island we were all down with it and the whole of the squadron ground crew were quarantined in a large marquee for ten days until it cleared.

The whole procedure of modifying the aircraft was once again carried out and on 8th November 1957 a megaton weapon was detonated off the end of Christmas Island. All personnel, except those required at the airfield, were taken by lorries to a large coconut plantation close to the main camp. We were dressed in bush hats, KD shirts, sleeves down and buttoned at the wrist, long KD trousers, socks and shoes. Sun glasses were compulsory. We were sat in the coconut plantation with our backs to the explosion area, hands covering eyes with sunglasses on. We could hear over the tannoy the aircraft carrying out its final run and the weapon release. There was then a count of 30 seconds – 20 – 10 – 9,8,7,6,5,4,3,2,1, FLASH. At that moment everything was quite surreal, there followed a most eerie silence – there was a huge brilliant flash of light, we could see every bone in our hands, at the same instant we experienced intense heat on our backs for approximately 3 seconds, the heat was then gone. It was as if a very strong heat lamp had passed over our backs, the immediate impression was that it was going to burn you, but then it passed very quickly. At this time there was absolutely no sound other than the tannoy intoning the countdown to minus 30 seconds and then informing us we could stand up, turn round and look at the fireball but we must keep our sunglasses on.

The sight was absolutely incredible – all the clouds in the surrounding area had vaporised leaving a bright blue sky, against which there was an immense single column of dark grey, black and deep crimson flames and fire rapidly rising and seething into the air. To get a better sight of this 'thing' a number of us scrambled out to the parked lorries and climbed aboard to enable us to see over the coconut palms. At this time there was still no sound but we were suddenly aware of palm trees in the near distance bending and swaying in great agitation. The next moment we experienced two huge explosions in rapid succession followed immediately by a powerful tornado-type wind. We toppled off the lorries more in shock and amazement than anything else as the shock wave passed. Nobody

Grapple X.

by the lorries was injured in any way, however, those who had stayed in the coconut plantation were showered with dislodged coconuts which fell from approximately 40 to 60ft – a number of the personnel were quite badly knocked about and in subsequent tests all observing personnel were seated out in the open on the beach in the port area of the island. Tank-landing barges were anchored by the shore with engines ticking over and would have been used to take personnel out to sea and safety should the wind change direction and blow the contaminated cloud over the island.

We remained for quite some time observing the huge column, a climbing, soaring, seething mass. It was still quite visible even at nightfall. It was then back to work at the airfield. On arrival at the airfield we found damage to aircraft and buildings. The scientists' hangar where aircraft were under cover and armed had one side caved in. Helicopters that were static on the airfield had windows blown out, and aircraft fuel storage tanks were also damaged.

From the point of view of the scientists Grapple X was a huge success as the intended yield had been estimated at approximately one megaton with a second test to follow, but to the surprise of everyone this first test produced a yield of 1.8 megatons. Therefore the second planned test was cancelled and the Valiants once again returned successfully to RAF Wittering.

However, the scientists were determined to test further theories and therefore Grapple Y was instigated which turned out to be Britain's biggest explosion yet. The test was carried out on 28th April 1958. I did not attend this test but I subsequently attended Grapple Z in August/September 1958 which entailed two more bomb tests from Valiant aircraft off Christmas Island and also two smaller tests from balloons. These were observed and witnessed by us from the beach across the bay near the port area.

During this visit to the island a tragedy occurred when one of our number died in his bed overnight of a heart attack. A very moving service was carried out at the station church followed by a burial at sea. This is the only time I have actually witnessed a sea burial and I must admit that it turned out to be rather grotesque. The small ship carrying the body, padre, bugler and crew were situated in deep water with the whole of our squadron and observers in barges forming a semi-circle a little distance away. After a short ceremony the bugler sounded The Last Post and the body, which was on a flat board covered in a canvas sheet with the Union Jack covering it all, was ceremoniously tipped down a ramp into the sea in sight of us all. This operation was of course intended to be the end of the sea burial. However, it was not to be because after a short while when we were preparing to return to shore the whole contraption suddenly appeared on the surface. The outline of the body could be plainly seen under the wet canvas and panic ensued as attempts were made to puncture the canvas and also to throw chains around it to sink it. This was finally achieved. That is an experience I never wish to see again ever.

Grapple Z concluded our test programme and on return to RAF Wittering the squadron resumed their normal role. I remained with the squadron until September 1959 when I was posted to Singapore for two and a half years with my newlywed wife, Rosemary.

Bill Evans' comments on the food contradicts sharply with Don Briggs. It is difficult to believe that the food the Hastings brought in from Hawaii wasn't shared out equally. I also asked Bill if there were any instructions about protection from radiation.

To answer your question if there was any talk of shielding us from radiation, the answer is absolutely nothing. In fact I often think how bizarre the whole of the tests were in some respects because we, the 49 Squadron personnel, were never briefed on what to expect or what we may experience in any of the tests from Buffalo to the Grapple series. The only order/advice we were given was to turn our backs to ground zero when the explosions were due. X and Z were quite something really, especially X which was rather frightening as we just did not know what to expect.

Clive Cox was an electrical fitter on XD824 and relates his account of Grapple X.

The first test carried out over Christmas Island was spectacular. The aiming point was a triangle at the furthest south-easterly point of the island, which was also the release point, however, there was a delay of about eight seconds between pushing the release button and the actual release. The most vivid recollection was when we were taken in lorries to a point at the north-east of the island. We disembarked, the PA system told us to sit on the ground with our backs to the point of burst, then to hunch forward covering our closed eyes with our hands and wait. Countdown began and at zero there was a flash so intensely bright that finger bones were clearly visible through closed eyes, at the same time a surge of heat swept over us.

After a few seconds of count up we were told we could stand, turn and look. It is difficult to describe the vision rising in the sky, it was a huge ball of mainly red, orange and yellow colours roiling and surging as it climbed ever upward. I then saw a white ring breaking outwards from the point of burst and as it struck clouds they vanished only to reappear afterwards. The ring came toward us at an alarming speed, I suddenly suspected it was the blast wave, when it passed over us there was an ear-shattering crack followed by a rumbling noise that I assumed was coming from the fire ball as it ascended. It was the most awe-inspiring experience of my life and frightened me somewhat at the truly terrible power of what I had seen. Some days later a group of us went to the aiming point, we found trees that were burnt black down the side facing burst point, sea birds dead with all their black feathers burnt away yet white ones remaining. Large pieces of coral were as light as a feather, the heat had dried all the moisture out.

Grapple Y on 28th April 1958 was a more efficient version of Grapple X with a yield of 3 megatons, the biggest bomb the UK ever exploded. Sqn Ldr Bates and crew in XD825 carried out the test. The height of the burst was 8,250 feet with a yield of 3.0 megatons.[11]

Grapple Z was a four-bomb series but only one was a bomb dropped from a Valiant on 11th September 1958; it was a smaller 1.2 megaton version of Grapple Y. Sqn Ldr Bailey and crew in XD822 carried out the test. The height of the burst was 9,240 feet with a yield of 1.0 megatons.

The tests were ended after Grapple Z as the Nuclear Test Ban Treaty came into force towards the end of 1958 and there was a new UK-US Mutual Defense Agreement. However the United States conducted a further 105 nuclear weapons tests at Christmas and Johnston Islands in 1962 and 1963 as part of Operation Dominic.

Operation Grapple was the largest British military operation since World War II, involving all three branches of the UK armed services (air, land, sea) with meteorological support from the New Zealand Navy. An estimated 22,000 servicemen from Britain, New Zealand and Fiji served during Operation Grapple and many were exposed to radiation from the explosions; some 1,200 civilian and military personnel were actually stationed on Christmas Island.

As a result of the tests, Britain claimed entry into the club of nuclear super-power nations, along with the United States and the Soviet Union.

The British Ministry of Defence (MOD) stated that the tests were

---

[11]   https://www.youtube.com/watch?v=pqCdxoItLbw

planned and conducted with 'meticulous care and the health and safety of test participants was of the utmost importance'. The MOD maintains there is no evidence of excess illness or mortality amongst the veterans due to the tests or exposure to radiation, and therefore no grounds for compensation but this view has been challenged, see Appendix One.

Canberra sampling Grapple. (*RAF Museum*)

Chapter Nine

RATOG TESTING

Milt Cottee was a member of the Royal Australian Air Force and completed the Empire Test Pilots School Course 14 in 1955 the year after me. By the time Milt was on the Squadron the Valiant was in service, the Suez campaign was over and the aircraft was carrying out nuclear weapon tests in the Pacific.

The Valiant operational clearances outstanding on B Squadron related to special equipment. This chapter is an extract from Milt's autobiography which covers his Valiant flying at Boscombe Down.

I joined B Squadron at Boscombe Down in January 1956. February was a busy month with Beverleys and Valiants, autopilot trials for the Beverley and rocket assisted take-off trials for the Valiant. On 28th February I took the NBS Valiant WZ373 to Idris in Libya for some bombing trials on the El Adem range; the tests were not particularly memorable but I remember the crew chief; he was a big red-headed Scottish flight sergeant. This fellow was known to like a binge now and then. The day before we were to leave Idris, he had a session in the NCO's mess and then decided to go into Tripoli.

Unable to find any suitable transport to Tripoli, which was about thirty miles away, he walked into the village at Idris and attempted to take over a truck from one of the locals. This caused so much consternation and yelling that the local constabulary were soon alerted. The crew chief was promptly arrested and thrown into a police cell. Some hours later, after a sleep and now somewhat less inebriated, he realised his predicament and thought it was about time to leave. Feigning illness, he somehow persuaded his jailer to enter his cell, whereupon the hapless fellow found himself rapidly changing places by dint of heavy physical persuasion.

At breakfast the next morning, I learned that our crew chief had the aircraft ready to leave early as the local police were demanding, through the base commander, an identification parade of all base personnel. Not wanting to leave my crew chief behind in Libya, I gathered the rest of the flight crew together and we expedited our departure, after getting the nod from the base commander.

The flight sergeant had filled our fuel tanks with more fuel than I had planned. On take-off we had a load classification number which exceeded the rated strength of the runway at Idris for landings. Soon after take-off, the entrance door seal blew out with a bang. I had a perfectly good excuse for not going back to Idris and instead diverted to the big USAF base at Wheelus Field about fifty miles away.

No one at Wheelus had ever seen a Valiant before, let alone one captained by an Australian. The USAF was able to help with repairs to the door seal, but not before I committed to staying overnight. After a meal in the officers' club that evening, we found some one-armed bandits and clubbed together to try our luck. Sqn Ldr Ackerman, the nav trials officer, was feeding quarters into a machine whilst I had my turn at pulling the handle. A resultant jackpot had piles of money pouring out of the machine on to the floor. This more than covered our personal trip expenses including those for the remainder of an enjoyable evening. Needless to say, the identification parade at Idris had not come up with a culprit so international relations did not suffer unduly.

The airfield at Idris had an avenue of large eucalyptus trees which had produced a lot of gum tips. I had picked a large bunch which I had secured in the weapons bay. They were a bit droopy by the time we landed at Boscombe and were promptly seized upon by the customs officer whose job it was to clear us back into England. Imagine trying to explain that the bunch of leaves were not some exotic new drug and were just a bit of nostalgia for an Australian family. He insisted on taking a sample with him and an intention to call someone in Australia House in London. I heard no more of the matter.

RATOG take-off before WB215 broke its spar.
(*Aviation Historian*)

April was not without its adrenalin-producing moments. Routine autopilot and bombing release trials were the norm and I began to concentrate on the Mk1 Vulcan. On 29th April 1957 I started up Valiant WB215 (the second prototype) fitted with two 4,000lb thrust Super Sprite rocket-assist units. The AUW was

170,000lb which was a bit over the normal AUW for a Valiant of 167,000lb. The weight was achieved using a tank containing 10,000lb of water in the weapons bay and two 1,600 gallon underwing tanks, also full of water. All of the water totalling 42,000lb was to be jettisoned after a measured take-off and the first releases of the Super Sprites after use. Flt Lt Ray Bray was my co-pilot and the only other crew member was the flight test observer, Steve Brown. We had lots of cameras and instrumentation as well as a closed circuit black and white TV camera showing us a 14-inch screen view of the left Super Sprite.

A Meteor 7 was to be our chase, with a photographer in the rear seat. The drop area was a field near Boscombe called Filedean. A ground party was there, ready to observe the releases, support the trial and recover the Super Sprites.

The take-off was routine and hardly spectacular at such heavy weight. My red-line speed for carriage of the rocket units was 230kts and the target drop speed was 215kts. I established a speed around 215kts while manoeuvring for a dry run over the dropping point at 1,000ft AGL. Smoke was released on the ground to guide my positioning for release.

Valiant WB215 lumbered around with the Meteor in right line abreast and all seemed to be going well. There was slight turbulence causing the aircraft structure to shake gently as I rolled over to about 20° of bank, turning onto a wide base leg. About 30° into the turn, there was a colossal bang. At the same time, a tremendous shock went through the aircraft. Roll increased to the right, this being instinctively corrected by aileron. I exclaimed over the intercom saying, "We've had a mid-air. The Meteor must have hit us. Be ready to get out." I called on the radio, concerned now about the Meteor crew, saying, "Splash are you all right?" Flt Lt 'Splash' Moreau came back laconically saying , "I'm OK – what's up?"

I had stopped the Valiant turning and looked for warning lights and any abnormalities, trying to get some clues as to the cause or results of an explosion. In the absence of any internal indications, I described to Splash that we had experienced a violent explosion and asked him to give us a good examination. Soon he said, "Hey, the speed brakes are extended slightly on the right wing and fully in on the left."

Meanwhile, we had a good look at the left Super Sprite unit on TV as I had thought that one of the units may not have purged its high test peroxide (HTP) properly after burn out, as it was supposed to do, and then exploded. It looked fine and Splash reported the other one as looking OK.

I checked the speed brake selector control and decided not to use it for the remainder of the flight. I also carefully exercised the flying controls. Failing any positive indications of any further problems, I decided to continue with the trial which involved jettisoning each Super Sprite singly on two runs over the drop zone. Instead of waiting to start dumping our heavy load of water after the releases, I initiated the water dump forthwith.

The release runs were routine and we were relieved to have the second one go away cleanly. By now, about half of the water had gone and we were rapidly

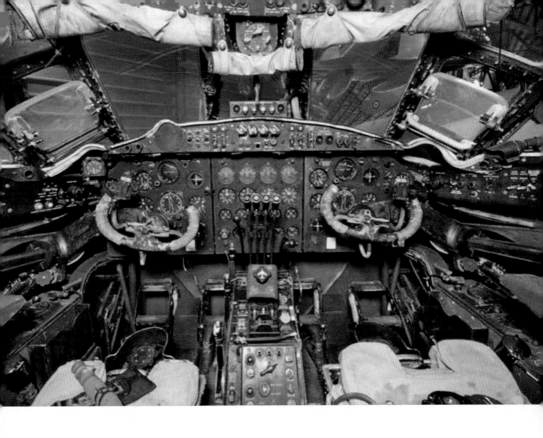

XD818 flight deck (above) and rear view (below). (*RAF Museum*)

Right: Poster promoting the 1958 SAC bombing competition. (*Roy Monk*)
Below: XD816 of 214 Squadron making a low level flypast at Wisley in May 1967. (via *Peter Sharp*)
Opposite top: Valiant taking off from Wisley. (via *Martyn Chorlton, Old Forge Publishing*)
Opposite bottom: Valiant WB215 before its spar broke during RATOG take-off. (*John Matthews*)

**TENTH ANNUAL
STRATEGIC AIR COMMAND
BOMBING-NAVIGATION COMPETITION**

RESC

October 13-18, 1958
MARCH AIR FORCE BASE, CALIF.
CASTLE AIR FORCE BASE, CALIF.

Above: Valiant XD818 at Christmas Island. (*Clive Cox*)

Right: Wg Cdr Hubbard entering XD818 before dropping the bomb. (*Clive Cox*)

Opposite top: Valiant with Table Mountain in the background. (*Arthur Creighton*)

Opposite middle: Valiant WP199 with Pegasus test engine for Hawker 1127. (*Aviation Historian*)

Opposite bottom: Valiant Blue Steel at Woodford, author at helm. (*Aviation Historian*)

Top: Painting depicting XD818 carrying the first H bomb. (produced by *Peter Sharp*)
Above left: 49 Squadron's headquarters at Christmas Island. (*Clive Cox*)

Above right: XD824 flight crew and ground crew. (*Clive Cox*)
Opposite top: XD824 ready to go. (*Clive Cox*)
Opposite bottom: Ground crew having breakfast on Christmas Island. (*Clive Cox*)

Top: XD822 crew returning from Christmas Island. (*Jo Lewis*)
Left: Valiant at Christmas Island. (*Clive Cox*)

becoming much lighter. I headed straight back for Boscombe Down with gear and flaps operating normally. Water dumping was completed on short final.

By the time I climbed out of the aircraft there was a group of airmen peering up under the right wing at a location just aft of the wheel well. An area of wing skin having a rough diameter of 18 inches had disappeared. I looked up into the hole and asked if anyone knew what was up in that area. It was mostly empty of anything significant. I had convinced myself that this had to be the location of our explosion so I asked that a stand be brought underneath and a torch obtained so that we could have a good look inside.

Meanwhile, a flight sergeant was more interested in a section of bottom spar cap which had been uncovered where some of the wing skin had torn away. It was obvious that rivets had been sheared off and that these were pointers to the source of the skin tears. Soon he was telling us, "The spar cap has a crack running through it". Positioning his ladder for a better look at the back wall of the undercarriage well, which was also the web of the main spar, he was able to add, "and the crack runs right up through the web".

The senior engineering officer was now present and he decided that the aircraft should be de-fuelled and then jacked up in a hangar using wing jacks. I was present about an hour later when the wing jacks started to lift the aircraft. It was with shocked disbelief that we saw the wing spar spread apart.

Meanwhile, Vickers at Weybridge had been alerted to a possible serious structural problem with second prototype Valiant WB215. As soon as they heard that we'd found a broken main spar, their senior engineers piled into a Heron aircraft to fly down to Boscombe to inspect the damage. Their initial reaction was that we must be mistaken. To them, a main spar failure would inevitably result in wing separation.

By now the full weight of the empty aircraft was being supported by the wing jacks. The spar cap had parted by about an inch and with the aid of a torch I could make out fatigue cracking tide marks over some of the break surfaces. I wanted a better look at the surfaces of the break and inquired whether anyone had a magnifying inspection mirror. No one had one so I decided to jump into my car and go to the base dental section to borrow one.

By this time, my adrenalin levels must have been fairly high. I had begun to comprehend how close I had been to oblivion. The implications to the whole Valiant fleet had also begun to become significant in my thinking. These were the thoughts I was having as I drove towards a most improbable location.

The appearance of an agitated pilot in a flying suit in the reception room of the dental section certainly got the attention of the receptionist and the waiting patients. There was an immediate ominous silence for a few seconds during which I grappled with the problem of how to explain that I urgently needed a magnifying dental mirror. To me, this was urgent but in retrospect there really was no such urgency about the situation.

I announced with much agitation that I needed to see the dental officer urgently and headed in the direction most likely to find him. On opening a door to the protestations of the WRAF receptionist, I charged into an occupied dental surgery, complete with a white coated dentist and a patient in the dental chair.

I blurted out: "We have just had a major structural failure in the wing of a Valiant aircraft. I wish to borrow one of your magnifying mirrors to assist with our preliminary investigations."

I picked up a mirror from his palette table to see if it could be the one I wanted. This problem had obviously never arisen in a dental section before. Neither the dentist nor his receptionist knew how to handle the situation. The patient was now sitting bolt upright in the chair. The mirror I had in my hand was non-magnifying. I asked the still incredulous dentist whether he had a magnifying mirror and I should take one of each. This seemed to trigger the start of the next scene. All the three others in the room started talking at once, each having a different question. I concentrated on what the dentist had said. He had asked whether the aircraft had crashed. I said, "No, but it was a near thing". The gravity of the situation had now been established and the dentist picked up a mirror and said, "Will this do?"

As I examined the mirror, the dentist asked, "May I come with you to see the aircraft?" It was now my turn to be incredulous and I laughed at the vision of me arriving back at the aircraft with the broken wing with a dentist in tow. It didn't occur to me to ask if he had an appropriate security clearance as I wondered how I could keep him at arm's length. But it was I who had set the scene and to his delight I assented.

The dentist quickly told the receptionist to look after things for a while and we climbed into the car as I introduced myself and gave him a brief description of the situation.

The engineers at the aircraft initially took the dentist to be a scientist and treated him with some reverence. But my commanding officer drew me aside and asked who he was. I quickly explained. He suppressed a laugh and took on a more serious face as he approached the dentist who was examining the fracture surface using his mirrors. The CO said in a voice which was on the point of breaking up, "What do you think sir?" The dentist stumbled out something to the effect that he thought it was very serious and perhaps he had better get back to his patient. The CO readily agreed and asked someone to take him away after thanking him for the loan of his mirrors.

The engineers from Vickers arrived soon afterwards and rapidly became as incredulous as the rest of us. The CO called for a conference in his office in half an hour. During this time, I had to take my turn for a closer examination of the break.

At the conference, the Vickers engineers stated that the wing must have come close to complete failure. Wing twist following the spar failure had been prevented and the load re-routed by a cross bracing steel rod which attached between the top of a forward sub-spar and the main spar bottom cap, outboard of the failure. This rod and some strength remaining in the wing skin had taken the

The three Vs flown from Boscombe Down with Milt flying the Vulcan.
(*John Matthews*)

load. Their estimates indicated that, had the aircraft exceeded about 230kts with any significant bank angle the wing would have totally failed.

Significant also was the fact that the main spar of the prototype had been initially designed for a lesser AUW than 167,000lb. In an attempt to bring WB215's spar up to production aircraft strength, a reinforcement had been fitted with additional metal strips being attached to the top flanges of the main bottom spar cap. This reinforcement had been discontinued at the very point where the failure occurred. Such an arrangement obviously caused a concentration of bending stress at that point which inevitably led to the growth of a fatigue crack. The dental mirrors had initially indicated that about 30 per cent of the spar cap cross section had failed in service by cracking and the remainder had failed in concentrated tension.

Valiant WB215 had made its last flight. It was used for a while for 'accelerate-stop' trials. It was established that the spars in the Valiant fleet were generally subject to fatigue cracking. A few years later, the Valiant was phased out of service. As of this date (July 1999) I believe that the three-man crew of WB215, during its last flight, are the only living survivors of an aircraft having suffered a main spar failure.

Another Valiant flight I remember was with Geoff Fletcher and involved a maximum rate stop on a partially wet runway. This was to determine the operation of the anti-skid units and the energy absorption capability of the brake disks. After touchdown, Geoff applied full and continuous brakes whilst holding straight with nose-wheel steering. The aircraft bucked about a bit and we felt two tyres blow out before we came to a lurching stop. The fire crew standing by had been briefed not to spray anything other than gas or powder on to the resultant smoking brake units. Water would have caused the brake disks to disintegrate and initiate secondary damage.

We left the aircraft on the runway where wheels were changed before it could be towed clear. I had a good look at the brake disks and pads afterwards. They had been much distressed but had succeeded in absorbing the intended levels of energy. All brake units had to be replaced.

Probably the most publicised flight was 14th September 1958 which was set aside for Battle of Britain activities. We were able to have the three V bombers, Valiant, Vulcan and Victor airborne together to enable some formation pictures to be taken from a Meteor. This was the first occasion that the three Vs were to be photographed together and the shots received much publicity. We then flew around the country for flypasts of most of the famous wartime fighter bases.

It is impossible to know what the Vickers design team made of the spar failure and how much hindsight Milt is attributing to the event. Milt clearly was not aware of WP217 in 1964 when the pilot, like Milt, was able to land the aircraft without the wing coming off. It is difficult to believe that Vickers did not see the writing on the wall for the Valiant at this stage bearing in mind the date and the international engineering warnings that were then available spelling out the dangers of using the alloy DTD683 for main spars. This point is discussed in more detail in Chapters Eighteen and Nineteen.

Chapter Ten

VALIANT MEMORIES

I was delighted when **Robby Robinson** agreed to do this piece for the book because he is an accomplished author writing about his extensive test flying experiences with a splendid light touch. I wondered if he would have anything to say which he had not already put down in his books but I need not have worried. We know each other well as I recruited him when he was at Boscombe Down to join my team at Avros.

In 1955 I was coming to the end of my tour on Canberras on 10 Squadron when my CO asked me if I would like to volunteer for the V Force, then just forming. I knew that the first V bomber was to be the Valiant and I was always one to want to fly the latest aircraft in the air force inventory. So, on 17th September 1955 I reported to 232 OCU RAF Gaydon to be told that I was to be trained as a captain and then to instruct on the Valiant simulator, the first in the service. I was mildly surprised because I knew that captains had to have a minimum of 1,750 hours, four-engine experience and a tour on jets. I only fulfilled the last requirement. In fact, I only had some 700 hours total flying.

Squadron Leader Stormy Gale, the chief simulator instructor, marched me in to see the chief flying instructor, Wing Commander Hank Iveson who promised to send me back to Honington to rejoin 10 Squadron but having just paid to move my caravan and wife to Gaydon I threw myself on his mercy and he kindly said that I could become that poor, benighted individual, the progress officer, who organised the ground school courses. Penury forced me to accept and I settled in to be the dogsbody to Wing Commander Deaves, the chief ground instructor. After a year or so I started to chafe at the paucity of flying I was able to do, so I added myself to the Bomber Command list of the next ground school course, the members of which were to form the

90 Squadron Valiant XD862 until it burst its tyres landing at Kinloss. (*via Martyn Chorlton*)

nucleus of a new Valiant squadron, number 90, to be based at Honington. I was
to be the co-pilot to Sqn Ldr Chris Spencer, six foot seven and nineteen stone,
the ex-Bomber Command heavyweight boxing champion, by no means a gentle
giant.

90 Squadron reformed in January 1957, but we were unable to collect our
aircraft from Vickers at Weybridge until the end of February. When we had
sufficient aircraft we began to work up to combat status being issued with our
first targets, which were to be studied on a regular basis. I cannot remember what
the first was but, like all our future targets, it was a military installation. There
were never any city targets. We now started to play our part as a deterrent force.
As our aircraft were all painted anti-flash white we called ourselves 'the great
white detergent'. Our C-in-C, Air Vice-Marshal Sir Kenneth (Bing) Cross, was
determined to match his USAF friend and rival General Curtis Le May in terms of
readiness. Compared to the American SAC we had a fraction of their numbers of
aircraft and crews but being closer to the USSR we were in a crucial position and
would be in the forefront of any retaliation. We could not copy SAC's ability to
have a standing force of bombers and tankers nor to have a shift system of crews
on standby. Therefore we were all on constant alert and could, and were, called
to readiness from bed or, on occasion, recalled from leave. This was done quite
regularly for station, Group or Command exercises.

I was a co-pilot for two and a half years before taking over the crew as captain
and, almost unnoticed by us, the constant stress began to tell. For some reason this
particularly affected the navigators and observers, perhaps because they bore the
brunt of target study. Some succumbed to ulcers and we had at least one nervous
breakdown on the squadron and, possibly, two suicides. Let me illustrate how
'twitched up' we could be.

All aircrew had to live on base. A tannoy system was installed and a chain
of loudspeakers was placed outside the married quarters. This system was for
operational use only. I developed the habit of parking my car pointing in the
direction of the operations centre and laying out my clothes beside our bed in
the right order to be quickly donned. One night my wife and I were sound asleep
when I woke with a start because I was sure that a tannoy announcement had
been made. I leapt from the bed, dressed and drove to the ops centre like a mad
thing. I ran into ops and shouted at the duty airman, "What's the H hour?", this
being the timing point for any exercise or for the real thing. The airman looked
startled and said that no announcement had been made. It turned out that all I
had heard was the 'click' when someone had accidently hit the switch. I felt a
fool but two other captains followed me in. Eventually even Bomber Command
recognised that something had to be done and over the next three years a proper
QRA system was established, with a proper crew roster. Also dispersal airfields
were designated with readiness hard standing at each end of the runway and crew
accommodation in caravans.

Much has been made of our ability to get four aircraft airborne in four minutes from the signal to scramble. This has gained some scepticism over the years but we could do it and practised it regularly. The drill went as follows.

Before going on standby the crew would go out to their aircraft and carry out all the checks as though they were going to fly. The engines would be started and shut down and the electrical power switched off. Helmets and other kit would be left in positions for quick donning and the crew would return to bed or the crew room. On the call to scramble the crew ran to the aircraft, where the ground crew had already plugged in the external power, and climbed the short ladder to the flight deck. The AEO would lead and switch on the batteries followed by the captain who would reach behind his seat and press the starter button for the first engine. As this engine wound up he would swing himself up into his seat and open the appropriate HP cock. He would select the next engine, press the starter button and start to strap in. He would repeat this and when all four were started he would release the brakes and power forward, the chocks and ground power plug having been removed earlier and the ground crew lying on the ground to avoid the jet blast. The aircraft was then turned onto the runway and with no radio clearance would take off. Easy! Although some problems did occur. During one night scramble one aircraft found itself in the air with no observer and another with six instead of five crew members. The drill was simplified in the later V bombers with the introduction of 'simstart' where all four engines could be started together.

Of course, life was not all grim and tense. We still found time for parties and we had regular detachments overseas. We went to Nairobi, Malta, Cyprus, Singapore and Butterworth in Malaysia. One very memorable trip was from Changi in Singapore to Manila in the Philippines and on to Saigon, Vietnam. This was in 1958 before the war there became really serious. There were two aircraft taking part, ours and the boss's, Wing Commander 'Freddie' Hazelwood. We led with me map reading from a very sparse map provided by Far East Command. I found the airfield and we carried out individual displays before landing to tumultuous applause from an uncontrollable crowd which surged forward and crowded into the cabins before all of us could leave them. Unknown to us we were the first jets ever to be allowed into the country, why we never did find out, but this accounted for the enthusiasm. We were taken to the officer's mess where drinks were pressed on us from a bar that was nearly six foot high; this, we were told, was because the French, whose mess it had originally been, did not want the locals serving them to see what went on. An extremely old trusty sat on a tall stool and handed down your drink. We stayed one night in that lovely city, visiting the night clubs, before leaving for Changi.

On one detachment to Cyprus we were inspected by the general in overall charge of the island's forces. The CO showed him round the aircraft and took him aboard the flight deck. We, the crews, stood to attention in front of our aircraft as the much-bemedalled army officer inspected us. As he departed he said to the CO,

"I suppose that you have to have very competent NCOs to fly these for you?" He was shocked when told that we officers had to fly them ourselves.

During the same detachment we took part in exercise Rosie Rosie, a NATO exercise that involved a simulated nuclear attack on the USN Sixth Fleet. The exercise was to be over four days. On Day One we were to lead the attack. The start time was to be 09.00 hours and we were ordered to cross the start line off the coast of Lebanon at that time. We took off at 07.00 hours and climbed to 40,000ft before turning to run in, crossing the line at a few minutes after 09.00. As we started our bombing run we could see the whole fleet laid out before us and could hear the chatter on the exercise frequency as the carriers tested their aircrafts' radios. Our observer, Don Richardson, took over steering the aircraft as he carried out the radar bombing run. He called "bomb gone" which I repeated over the radio. The whole fleet went silent with palpable shock and then a lone voice said, "Aircraft saying bomb gone identify yourself". We had caught them with their pants down. Apparently they protested at high level that we had cheated. Nevertheless, they were on the ball for the next three days. All's fair in love and war.

The squadron remained in the main bomber force until the end of 1961 when we were told that we were to become a tanker squadron joining 214 Squadron as the RAF's first tanker force. The news was received with joy. No more standbys, no more target study. We started to be trained by 214 Squadron in the art of tanking, which was quite easy, and receiving, which was difficult. In the first week of training I was told that I had passed the interviews and exams for the Empire Test Pilots School. Bomber Command decided that they would save money by not training me to receive. The Ministry of Technology, which ETPS came under, told them that it was essential that I be receiver qualified for my future employment at Boscombe Down. Bomber Command responded by giving me only two hours of training and not the ten hours it usually took. I never did get any good at this art.

My last serious Valiant flying was to refuel the CO's, Wg Cdr Ulf Burberry, aircraft over Goose Bay as he carried a foundation stone for the cathedral in Chicago from the ruins of the abbey of Bury St Edmunds. My crew and I spent a week at the Goose arranging the rendezvous and waiting for the signal to go. On the due day we sat in the aircraft in our arctic clothing, but still freezing, looking out at the snow-swept pan where the poor ground crew cowered against the forty-knot wind. Stan Mellor, our AEO, announced that he was in contact with the incoming aircraft and we started up and taxied out in a total blizzard. The take-off was hairy to say the least but we knew it had to be done as the boss had said that he was short of fuel and in that weather he would be unable to divert into the Goose. We intended to keep enough fuel for ourselves to divert into Sept Isle. We climbed to the agreed height and entered a race track pattern. The boss reported that he was at our position but could not see us. This was before the days of the tanker having an on-board tactical air navigation system (TACAN) beacon so I told Peter Diggance, the co-pilot, to jettison a small amount of fuel.

The boss saw this and closed on us. We trailed our hose and he immediately made contact and took on board a full load. He broke contact and thanked us in his usual laconic way before continuing on his way to Chicago. We transferred to the Goose frequency to say goodbye but were told that the weather had cleared so we landed there to spend another night in the snow-bound mess before flying home.

My very last Valiant flight in Bomber Command was on 20th January 1962 and was to deliver XD863, the aircraft we had collected from Wisley in 1957, to Marshalls at Cambridge to be converted to a tanker. A fitting end, I thought, to flying my favourite big aircraft. However, I was to fly the Valiant a few more times at Boscombe Down in 1963 before its fatigue troubles grounded it for good.

I related to Robby's remark on flight refuelling. Unlike him I never had any training in receiving fuel and I remember going to Boscombe Down with the first Vulcan equipped for flight refuelling as a receiver, Vulcan Mk1 478. My log book shows I did three flights with Pete Bardon, the first trip being dry contacts and the second two receiving fuel. I can't remember how I did but I do remember it was not a success story except that the aircraft worked OK.

Chapter Eleven

148 SQUADRON MEMORIES

232 OCU Valiants at Gaydon. (*Martyn Chorlton*)

Graeme Kerr joined the RAF as a radio apprentice in 1951 and won a cadetship to RAF College, Cranwell in 1954. On graduation from Cranwell, he did a Valiant conversion course at 232 OCU RAF Gaydon as a co-pilot and then joined 90 Squadron. After three years he got his command on Valiants at RAF Marham on 148 Squadron. Before leaving the RAF Graeme became involved in introducing the Buccaneer in Germany with UK nuclear weapons as distinct from US ones. He then flew Viscounts and the BAC 1-11 round Scotland with British Airways.

My first posting after flying training and also having completed a Valiant course at 232 OCU, RAF Gaydon, was to 90 Squadron at RAF Honington in early 1958. I should mention that the selection procedure for the V Force in 3 Group entailed an interview with the AOC, AVM Cross, with pertinent questions such as to whether one was a regular church attender, played rugger, and had a full mess kit; all questions had to be answered in the affirmative!

RAF Honington was home to two Valiant bomber squadrons, 7 and 90, and a further squadron of ECM Valiants on 199 Squadron. The latter moved to RAF Finningley and changed its squadron number to 18 and was replaced in time by a Victor squadron, 57 Squadron.

Bomber Command requirements for co-pilot selection had changed in 1958 and I was amongst the first to be selected without having had previous operational flying experience. Most of the crews were fairly senior, many with wartime experience, so young co-pilots were a novelty. The so-called young captains on 90 Squadron at the time, Flt Lts Cochrane and Goodall, had both previously done full tours on Canberra squadrons.

Co-pilots were not considered part of the integrated crew, which in time, with navigational and bombing results, was expected to progress from combat, to select and the ultimate select star rating. This was very important in assessing a crew's capability and also helped in selection towards the RAF team in the prestigious USAF SAC's navigation and bombing competition held annually in the US. Nevertheless, although changing a co-pilot did not affect the crew's operational assessment, crew integration was a major factor and in this regard co-pilots were included.

However not being considered an essential component of a crew, from an operational standpoint, had the advantage that one flew with many crews on the squadron and this meant that some captains gave co-pilots more handling than others. The lack of handling an aircraft, especially for first tour co-pilots, was something which obviously worried the authorities and as a result, Chipmunks were obtained on each station so that we could get some flying hours to build up experience.

Overall, my three years with 90 Squadron, initially under Wg Cdr Freddie Hazlewood, was a very happy time with many detachments to Malta and Butterworth in Malaya, together with lone ranger flights to Africa, the Gulf area and across the Atlantic to Canada and the US, particularly SAC Headquarters at Offutt AFB Nebraska.

A final story from this period. The RAF detachment at Offutt AFB was commanded by a squadron leader and consisted of a flight sergeant and about twenty airmen. Their purpose was to service and turn round Bomber Command and Transport Command aircraft transiting or visiting. The particular squadron leader in 1960 had committed a cardinal sin by misreading a signal and had stood the detachment down on the day the C-in-C Bomber Command was visiting. Thus he was required at HQ soonest to be told of his incompetence. He was Valiant qualified, so I was flown out to take temporary command of the detachment and he returned to the UK in my co-pilot seat. An abiding memory from that date is standing beside a USAF colonel, as we watched four Vulcans, in transit to California, do a 30-second stream take-off and near vertical climb – "... and these are British bombers?" he asked with incredulity.

I got my command on Valiants after three years in 1961 and had the dubious distinction of being the first 'straight through' and youngest captain, without previous command experience. This accolade did not last long, as six weeks later some even younger upstart on the Vulcan force had progressed in a similar fashion.

I was posted to 148 Squadron at RAF Marham (popularly called 'El Adem with grass'). There were two other bomber squadrons at Marham – 49 and 207 – and a Valiant refuelling tanker squadron, 214. My overall memories of my three years at Marham, until June 1964, are all together of a more serious nature. This is probably because the three bomber squadrons were committed to NATO with US nuclear weapons and each squadron maintained a crew in the QRA area, each for 24 hours, on fifteen minutes readiness. Frequent call outs to five minutes readiness were made by HQ Bomber Command, which involved getting into the aircraft, all checks done and ready to start engines. These call outs invariably happened at either 3am or just when sitting down to the evening meal.

I think it was Easter 1962, when the Campaign for Nuclear Disarmament (CND) movement was at its height, that they decided to demonstrate at Marham. As long as they stayed outside the base, there was no problem, but some of the more fervent members had threatened to break into the base and invade the QRA and bomb dump areas. This gave the authorities a major headache because a break in of this sort would probably have been met by firepower from the US guards, who would have had no qualms about shooting anyone not entitled to be in those areas. I think just about every RAF policeman in the UK was drafted in to Marham as a precaution.

To maintain the fifteen-minute, dropping to five-minute readiness states, it was obviously not possible to meet this target with more than a handful of aircraft airborne in five minutes. As a result, each squadron had two dispersal fields to which they would deploy at times of heightening tension. In the case of 148 Squadron our dispersal fields were Manston in Kent and Tarrant Rushton in Dorset. The latter was in a very rural setting, owned by Sir Alan Cobham, who also owned a firm called Flight Refuelling Ltd. From memory, Sir Alan got his runway resurfaced by Air Ministry, to meet the Valiant requirement, so he presumably had some friends there.

In general, accidents aren't something which aircrew dwelled on, and thankfully there weren't many during my time on Valiants. Two however do linger in my memory. One was at the aforementioned USAF base at Offutt, when a crew from 543 Squadron at Wyton failed to reach flying speed and ran off the end of the runway in freezing conditions. Amazingly the aircraft stayed upright, and slithered to a halt in farmland without thankfully any sign of fire. Total damage, one very sick Valiant and a navigator with a broken ankle.

The other accident at Marham was more devastating as it happened just after take-off and the crew, including a crew chief, from 214 Squadron, were instantly killed. The tailplane trim was subsequently found to be faulty in that it operated in the wrong sense and once selected, ran the trim to the full extent of its travel. Amongst the sustenance for long trips were tins of soup and heaters for these were installed in the aircraft. It was subsequently found that over time, soup had dropped on to the trim switches on the control column, and this had somehow reversed the polarity of the switches.

My second tour on Valiants ended in June 1964 and my scheduled ground tour was to RAF Gaydon as station adjutant. In the days when one was required to get flying hours, even in a ground posting, to maintain one's flying pay, this was ideal for me — I could still get my hands on a Valiant occasionally. Unfortunately this situation did not last long, as after one Flt Lt Foreman landed in an aircraft in which he had heard a loud cracking sound, this was the beginning of the end. Twenty-five per cent of the fleet were found to have incipient main spar cracks, probably due to having switched to operating at low level, where the fatigue level was much higher and for which the aircraft was never designed. The decision was to scrap the fleet in early 1965, with only one airframe surviving. It initially stood at the entrance to RAF Marham and through time has moved to the RAF Museum at Hendon and finally the Cold War display at RAF Cosford.

My Valiant memories were revived quite recently when I was visiting a friend in Rabat, Malta, who showed me a picture of not one, but three Valiants on the very short civil airfield,Takali. I was rather surprised as whenever we were visiting Malta we always used the military airfield at Luqa. I did some detective work and courtesy of Mr Hermann Buttiigieg who runs www.aviationmalta.com. I found that on 27th August 1960 there was an air show scheduled at Takali celebrating the 20th Anniversary of the Battle of Britain and the aircraft had pre-positioned the day before. Luckily one of the aircraft captains remembers the event:

"We operated from Takali, just below the old city of Malta, adjacent to Rabat. We were parked on the old fighter operational readiness platform (ORP) at the end of the longest runway, some 4,800 ft, 1,483m or so long; this was a bit on the short side, even for a Valiant. There were three of us – Sqn Ldr Pete Coventry, Flt Lt Mike Cawsey and myself and we all had very light fuel loads. We performed our normal Valiant scramble and on the day we had all three aircraft rolling at the same time. Each had a different role to play before the crowds. I can't recall what the other two did but mine was a very steep climb away at minimum speed, not much above stalling speed! The Valiant could match the climb-away attitude of the Vulcan, but with less power the rate of climb was not as good. On this occasion I climbed to 2,000ft plus, did a quick wingover and descended straight onto final approach at Luqa. My flight time was exactly two minutes and thirty-five seconds so rounding up it is in my log book as five minutes. Had I been able to cut the time by six seconds the flight would not have been loggable! Previously we had had a lot of fun during rehearsals. Pete Coventry led us in a tail chase all round Malta and Gozo with run-ins and breaks at Takali. We also managed to so some illegal close formation flying out of sight of the 'big wheels' in Malta."

I believe the Valiant is the most forgotten of the three V bombers. It didn't have the revolutionary airframe designs of the Vulcan and Victor, and it used the basic Rolls-Royce Avon engine which also powered the Canberra. It handled very much like an overgrown Canberra and had the added advantage that in the event of a total power failure, the aircraft could be flown and landed safely in manual control. I think the Valiant benefited from the somewhat delayed introduction of the later marks of Vulcan and Victor, in that there were many overseas trips – flying the flag I believe it was called – which would later go to the other two Vs. More seriously it comprised the major part of the UK's commitment to NATO, for quite a few years, as a heavy bomber with a range and bomb load that could operate well behind the Iron Curtain.

Chapter Twelve

543 SQUADRON MEMORIES

Gordon Dyer was first navigator on PR7 Canberras at Wyton and then he was posted to Gaydon and managed to get himself on a Valiant B1 course. However, he was then posted back to Wyton on photo reconnaissance, but this time with the Valiants B(PR1)543 Squadron. The following is an excerpt from his autobiography.

The period was also busy for getting ready for Operation Bafford. This was to be a ten-week detachment to Townsville, Queensland, Australia. The task was the survey of the Solomon Islands, New Guinea and the New Hebrides. Two weeks before we left I was told to go to the MOD in London and report to Deputy Director Ops Recce RAF. I was briefed there on a special mission which would involve taking some intelligence photography on the route out. The actual targets were so sensitive that I was told that I could tell no other person at Wyton, though clearly I would need at some time to tell John Coltman, our pilot/captain, that I would be asking him to change altitude and possibly direction at some point.

The sortie went without a problem and when we finally arrived in Townsville on 28th June 1962 the magazines were handed over to the detachment RE Survey officer who was with us, for special handling. The route also sticks in my memory because of the treatment we were given by the customs officers at Darwin. On arrival at an extremely hot Darwin the customs officer came out to the dispersal and instructed us to take everything out of the aircraft. This included not only our own cases but all the spares equipment that we were carrying in the camera (bomb) bay. This was absolutely packed with heavy spares and it took the six of us (we had a crew chief with us) about an hour to do. When the customs officer came out again he did not inspect the kit but simply said, "OK, you can put it back again now". This treatment after a long flight from Singapore left us very disgruntled.

The RAAF at Townsville could not have been friendlier, they gave us a great party on arrival and we soon settled in as full members of the mess. Our liaison officer, nicknamed 'Blue' (nearly everyone seemed to be called Blue) was great fun, and over the ten weeks he arranged many a social event for us. Perhaps the most memorable was an RAF v RAAF cricket match, a match quite different to any I had played before. It was based on levelling out any significant skill advantage that any player had in either team, by attempting to limit their effectiveness through alcohol. Bowlers were allowed to bowl only a limited number of overs. Any success by a bowler i.e. taking a wicket, was immediately celebrated by him having to drink another glass of beer. And, if the batsman's departure had involved another person e.g. by being caught or stumped then those involved also had to drink. Similar constraints were placed on batsmen, in the scoring of a boundary, or accumulating 10 runs, or 20 runs, or 30 runs which also brought the requirement to drink a glass of beer. A game of this type soon deteriorated into laughter and chaos, with both sides declaring victory and malpractice by the other side. But it was enormous fun and really cemented relationships between the Townsville officers and the detachment crews.

It soon became clear to us that the Australian male – and I suspect that this also applies widely outside of Queensland – had a very different set of priorities to that of the typical English male. I think I can best summarise this as being in the following order: first, drink; then cricket; next, fishing; then, roo (kangaroo) shooting; then, alligator shooting; and a long way after all these came the opposite sex. The latter became clear when on several occasions Blue decided that some feminine company would be good to liven up an event, he simply rang what he called 'the bulk store'. This was the term for the Townsville nurses training hospital and within an hour a bus carrying young nurses would arrive at the mess. The emphasis on beer as a top priority was a lot to do with the working day (which typically started at 7am and went to 3pm) and the drinking laws in Australia. The bars downtown stopped serving beers at 4pm and shut for several hours before reopening later in the evening. If a party of men went into a bar together they each bought a round on entry, so drinking carried on long after the bar shut.

Flying itself was a great adventure and took me to places that before I could only have ever dreamed about, namely to Fiji and Papua New Guinea (PNG). Our main target areas were the Solomon Islands, with secondary targets the New Hebrides and New Guinea. The latter were only secondary in the sense that the weather factor was normally very poor over these mountainous territories. The nearest part of the Solomon Islands, a chain of five islands running west to east, was some 900 miles away. This meant a typical 2hr 30 min flight to get to our first potential target. Timing of arrival in the target area was critical and had to be based on sufficient sun to remove any mist in jungle valleys but not with the sun being so high so that it would trigger cumulus cloud. This meant a local sun time of a start around 0800hrs. Due to longitude difference of some 15° (i.e. a sun time earlier than Townsville, and a latitude difference of some 12°) the flight meant a take-off at 0400hrs local. For us it meant a call at 0200hrs (which was hardly worth going to bed for) and for the local residents a loud roar of jet engines every morning at 0400hrs.

Our first operational sortie was on 4th July and as it turned out it was the most successful, as I managed to complete 477 nautical miles of flight lines in an overall flight time of 7 hr and 10 min. This was mostly over the island of Santa Isabel which I managed to cover completely. As it happened the flight map I was using was based on World War II maps used by the Japanese. It was clear to me as I looked down that the island was wider than the map indicated and so as a precaution I flew an extra flight line down the middle of the island. Just as well too, as the subsequent plotting of the prints showed that this was necessary.

The rules that we operated under were that if the weather was good and allowed for continuing survey, we could if we were at the western end of the Solomons, night stop at Port Moresby (PNG), or if at the eastern end of the island chain, night stop in Fiji. This allowed for a magazine change and then further survey on the way back to Townsville. We were fortunate in having stopovers on three occasions in Port Moresby, and an extended stopover in Fiji.

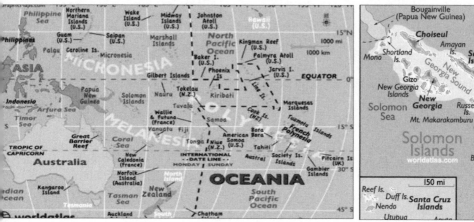

Map of Oceania/ Map of Solomon Islands.

Port Moresby was a bit of a paradise then. But it only had a comparatively short runway of 6,000ft. With the prospect of high temperatures during the day it was important to take off early, so it was not the place for parties into the small hours, but it was a great place for sightseeing. On our first visit the local British consul arranged a trip for us to visit a long house in the local village, and he laid on a reception in his residence where we met several of the local white community. Our aircraft was guarded overnight by the local militia. These endearing looking natives looked quite comical as the guns they had were almost as tall as them. When we arrived the next morning we were surprised to see that someone must have bribed them to look the other way as we found that a large image of a kangaroo (a symbol of the RAAF) had been painted on the tail fin.

Our stay in Fiji was also a delightful experience. We arrived on a Friday afternoon after survey work over New Hebrides. We had a planned stop at the Nandi Airport hotel and we were due for another sortie the next day. However, next morning John Coltman decided during his pre-flight inspection that a problem that he detected with the aircraft battery was too risky to proceed with and declared the aircraft unserviceable. This meant signaling Townsville and requesting a new battery to be flown to us. This would not be due till late on the Sunday at the earliest.

We had hoped that our accommodation at the airport hotel could be extended for another two nights. However, on returning to the hotel, the manager explained that they were full for the weekend and that we would have to go elsewhere. He said that he would phone around and make arrangements for us. So we settled into the luxurious seats in the hotel lounge and ordered coffee. After an hour or so the manager returned and said, "You're all set, your aircraft is waiting for you!" It transpired that he had found accommodation for us at the Korolevu Beach hotel, which was only accessible by air.

So we were taken in threes in a Piper Caribbean on a twenty-minute flight low over the jungle to our destination. The airstrip was grass and was approached from the sea on to what was a ribbon of available landing area by the side of a stream. When we stopped and checked in we discovered that the hotel was often used by honeymooners. The accommodation looked like native huts, called bures, but were in fact superbly fitted with every mod-con inside. Our enjoyment did not end there as, because it was a Saturday night there was entertainment laid on. This meant that we sat on the ground (on banana leaves) around a large fire and were brought food and kava drink while we watched local Fijian natives do fire dancing. Unfortunately we were only there for one night as the hotel was full for the Sunday night. We returned to Nandi and had a night in a third hotel, which though comfortable enough did not have the same draw as the Korolevu Beach.

The flying was also accompanied by the opportunity to get to know the area and enjoy off-duty experiences. This included a trip in a glass-bottomed boat to see the fish on the barrier reef. I also went on a boat to Magnetic Island, just off the coast. This had been the place where lepers used to be sent. I also got to know a local garage owner, called Greg, quite well. One weekend I went fishing with him along a river bank. After a while I had a bite on my rod and after a struggle I landed the fish ashore. As I looked down at it I saw that it was very ugly and was about to pick it up. But suddenly I was aware that Greg was screaming at me. "Don't touch it, it's a stonefish, and very poisonous." In a trice he plunged his knife into the back of the fish and pushed me away from it.

In between flying days it was also a good opportunity to practice at playing bridge which was a real favourite for all the crew. Tony Neal, our nav radar, and I had several challenge matches with a couple of Australian officers, and I am pleased to say, we usually won. A highlight of the trip was having a free weekend and catching a flight down to Sydney where I met my Uncle John. This was an emotional moment for both of us. Altogether the detachment to Townsville was very satisfying experience. It included the longest flight I had had to date in a Valiant of 8 hours and 10 minutes.

Many amazing sorties and altogether we flew 138 hours in the ten weeks away, arriving back in the UK on 29th August. I achieved a few small but significant ambitions. Nandi was at 178° of longitude east so we had taken the opportunity to fly a little east before returning to Townsville so that we crossed the dateline. On night sorties I had seen Canopus the brightest star in the southern hemisphere. Townsville was at 19° south so I had seen that the bathwater really did empty by swirling anti-clockwise.

When Gordon Dyer left the RAF he had an academic career with the Open University becoming a technology staff tutor and then member of the directorate.

Chapter Thirteen

214 SQUADRON TANKING

Shaun Broaders joined 214 Squadron in 1959 and spent the next five years honing the Valiant's flight-refuelling capability and preparing the way for the Victor and Vulcan refuelling squadrons. This splendid account includes photographs from the Cobham Archives and quotations from *In Cobhams' Company* by Colin Cruddas.

"You are posted to RAF Marham". The year is 1959 and fresh out of armament mechanic training the clerk advises me of my posting; "Report to the general office on arrival". I am posted to a bomber unit with the first of the V bombers, the Valiant. Leaving Melksham for Marham all kinds of visions pass across the mind, but you have to feel that a posting to a squadron would be much better, as it would be more exciting and a chance to see the world. On arrival at Marham it took a few months to settle in and attend the Valiant ground servicing course at RAF Gaydon and the Martin Baker ejection seat course Type 3A at Higher Denham and then it happened, I was posted to 214 Squadron, what joy.

Just to put you in the picture, concerning 214 Squadron and Marham, I think a little history would not go amiss. On 21st January 1956, 214 Squadron re-formed at Marham with Valiant B1s, first of the V bombers. Powered by four Avon 204 axial-flow turbojets rated at 10,050lb thrust. The advertised estimated performance:

Valiant before the wing tanks were fitted.
(*Shaun Broaders*)

speed 610mph at 40,000ft, cruising speed 530mph, service ceiling 50,000ft, initial rate of climb 4,500ft/min and range 3,000 plus miles.The same year 207 Squadron disbanded on 27th March and re-formed at Marham on 1st April with Valiants. 148 Squadron also re-formed at Marham on 1st July with Valiant aircraft. Then during September/October 1956 all three squadrons were detached to Luqa airfield in Malta to take part in the Suez operations.

In October, 214 Squadron carried out raids with 1,000lb bombs on El Adem airfield, Almaza and Abu Sueir, and in November the targets were Kasfarit, Huckstep Barracks and El Agami. All of these squadrons returned to Marham in November/December 1956. Also, with the arrival of the nuclear capability in 1957, RAF Marham was awarded its own station crest, a blue bull with the motto 'DETER'.

In-flight refuelling was not a new concept in 1956, history records that as early as 1919 some people were beginning to think that by transferring fuel from one aircraft to another in flight, thereby extending their flight times, this could be used for trans-continental flights and many combinations of aircraft have been used through the years trying to prove the benefits of in-flight refuelling.

Flight Refuelling Ltd (FRL) has had a long association with the RAF but the company was about to end its association with propeller-driven aircraft. In May 1953, following a decision to equip the RAF's V bomber force for aerial refuelling, Canberra B2 WH 734 arrived at FRL and after installation of the prototype Mk 16 hose drum unit (HDU), then developed for the Valiant tankers, it became Britain's first jet tanker.

Further studies of in-flight refuelling led to trials of a modified Valiant bomber, which started flying from Boscombe Down in 1956 and after clearance for service was obtained from the Aircraft and Armament Experimental Establishment (A&AEE), 214 Squadron began Valiant-to-Valiant flight-refuelling trials in 1957, which were in the main dry contacts. Then in March 1958, whilst retaining its bombing role, 214 Squadron became the trials and development unit for flight-refuelling trials number 306 and 306A. Trial 306 was to test the capability of aircraft tanker and receiver equipment, and Trial 306A was for developing modern rendezvous (RV) procedures and techniques; these trials lasted until May 1960. The initial training of both air and ground crews was carried out at Flight Refuelling Ltd at Tarrant Rushton, while the Flight Refuelling School was developed at Marham – and all subsequent training for flight refuelling in the Royal Air Force was to be carried out at this school.

Conversion of the Valiant bomber entailed the fitting of a probe to the front of the NBS scanner bay and connecting it internally to the aircraft fuel system and installing a Mark 16 HDU internally in the rear of the bomb bay and a 4,500lb fuel tank in the front of the bomb bay. The HDU control panel was positioned beside the navigator radar who now had the extra job as the fuel panel operator. External floodlights were fitted to give the aircraft the capability of flight refuelling at night.

The modification to the Valiant bomber gave the aircraft in the tanker role the facility of transferring 45,000lb of fuel at a maximum rate of 4,000lb/min with a maximum drogue fuel pressure of 50 pounds per square inch. Although there were early teething problems, these were overcome, and in January 1959 two fully modified tankers began wet transfers.

The 214 Squadron Valiants actually gave their first public demonstration of AAR at the 1958 Society of British Aircraft Constructors (SBAC) Flying Display, after which, during the Trial No 306 they were involved in many long distance proving flights and various interesting unofficial records were made.

As recalled in *In Cobhams' Company* by Colin Cruddas, Sir Michael was particularly inspired to make a two-way, record-breaking flight from England to South Africa in 1959, following Sir Alan Cobham's journey to the Cape thirty-four

HDU installed in the bomb bay; winding
in the hose.

Wg Cdr Michael Beetham (left) and crew of the
Valiant, which undertook the first non-stop flight to
Cape Town, 1959. (*Shaun Broaders*)

From the left, Pat Hornbridge, Dickie Dickenson
and Sir Alan Cobham visit 214 Squadron, Marham,
10th June 1959. With Sqn Ldr Garstin, Wg Cdr
Michael Beetham and Gp Capt Wilf Burnett. (*Shaun
Broaders*)

years earlier. His Valiant was the first aircraft to fly to Cape Town and back, non-stop in each direction, being refuelled by two Valiant tankers over Kano, Nigeria on both flights.

By now both aircrew and ground crew were enjoying flight refuelling. The main challenge to the aircrew was the exacting task of learning to fly as receivers whereas the ground crew had many opportunities for visiting exotic places such as Malta, Cyprus, Bahrain, Karachi, Mauripur, Nairobi, RAF Gan, RAAF Butterworth, Singapore and RAAF Darwin. Of course they also enjoyed flying in the Bristol Britannia, which usually stayed with the squadron throughout the detachment. You can imagine the enthusiasm when the next detachment was mentioned; almost all the ground crew personnel were quick to volunteer for the trip, knowing all too well that some were going to be disappointed.

May 1960 saw the completion of Trials 306 and 306A and also heralded the arrival of Wg Cdr P G Hill the new squadron CO. The squadron were kept busy with training and deployments overseas as well as taking part in trials with the Royal Navy Scimitar and Sea Vixen fighters and a compatibility exercise with USAF Destroyer, Super

WZ390 and WZ376 during trials. (*Shaun Broaders*)

Sabre and Voodoo aircraft. Other long distance flights during this period include:

a. Marham to Offutt 4,336 miles in 9 hr 30 mins 19 Jan 1960

b. Offutt to St Mawgan 4,400 miles in 9 hr 3 mins 25 Jan 1960

c. Marham to Changi, 8,110 miles in 15 hr 35 mins 25 May 1960

d. Butterworth to Marham, 7,700 miles in 16 hr 16 mins 1 June 1960

e. Marham to Vancouver, 5,007 miles in 10 hr 28 mins 5 July 1960

f. Vancouver to Marham, 5,007 miles in 9 hr 35 mins 8 July 1960

Another interesting exercise in which the squadron took part was the refuelling of a Scimitar aircraft from the Royal Naval Air Station (RNAS) at Lossiemouth to the aircraft carrier HMS *Ark Royal* 'somewhere in the Mediterranean'. We did find it!

The Javelin crews of 23 Squadron from RAF Coltishall started receiver training in 1960 with practice flights to Akrotiri and back in August, then in October four Javelins were flight-refuelled from the UK to RAAF Butterworth in Malaysia via Akrotiri, Bahrain, Mauripur, and Gan.

It was on this trip that a number of the squadron ground crew suffered the usual tummy upset at Mauripur, but fortunately we had the Javelin squadron medical officer with us. To ensure we did not drink any more water, I recall him giving authorisation for the squadron to obtain soft drinks from a local distribution firm for the duration of the detachment. This ensured that we all made it to Butterworth without further mishap. Ernie Hill, the squadron sergeant electrician, was very ill at this time and it was doubtful whether or not he would make it to Butterworth. However, with good nursing and plenty of fizzy drinks, he recovered enough for the onward trip to Gan and after medical treatment carried on to RAAF Butterworth with the squadron.

It was about December of 1960 that Vulcan crews started receiver training for the proposed non-stop flight to Sydney in Australia. Training progressed with non-stop flights from Scampton to Nairobi, and to Karachi and back then finally in

Refuelling compatibility trials with a Sea Vixen. (*Shaun Broaders*)

Valiant tanker of 214 Squadron refuelling a Javelin of
23 Squadron. (*Shaun Broaders*)

June 1961 with the first non-stop flight from Scampton to Sydney, a distance of 11,500 miles in 20 hours 3 minutes at an average speed of 573 mph; the aircraft was refuelled overhead Cyprus, Karachi and Singapore. If memory serves me right we had nine Valiants on this trip and all remained serviceable for the duration of the exercise.

After this significant achievement Sir Alan Cobham presented trophies to the officers commanding of 214 and 617 Squadrons. It was also memorable in that the double 'Speed Bird' logo used by FRL was now incorporated in the 214 Squadron emblem emblazoned on the tail fins of the tankers.

Other memories are of the boss, during one of the many training trips to Nairobi, coming home with a bomb bay pannier full of fruit for the squadron personnel. My wife and I recently visited Kenya and arrived at Nairobi airport only to find the aircraft parked opposite the RAF detachment office that we used, still proudly displaying the RAF sign. The bomb bay pannier was also handy for the Christmas run to Malta, when orders were taken for the Christmas bottle.

It should be noted for the squadron enthusiast that on 1st April 1962, 44 years after the formation of the RAF, 214 (Federated Malay States) Squadron and 90 Squadron officially became tanker squadrons, losing their bomber commitment and so becoming the first tanker squadrons in the RAF. HQ 3 Group retained operational control of these squadrons.

It was during the loading of a HDU to one of our aircraft that we nearly had a serious accident, although at the time we all fell about laughing. I used to drive and operate the Simons bomb hoist while Corporal Blower would get in the cage and I would manoeuvre him onto the spine of the aircraft and with the chief in the bomb bay we would carry out the loading procedure, during which time we would be in contact through a throat mike head set. After the load was completed it was standard operating procedure for the corporal to stow the equipment, remove his headset and throat mike and before placing them in the cage, hold the mike to his throat so as give me the instruction to withdraw the boom and remove the vehicle, (you must bear in mind that you could not see the top of the aircraft from the vehicle). He would then come down via the aircraft rear hatch. The instruction given I started to swing the boom, only to hear a strange noise in my headset; that's right, he was still attached and I was pulling him along the spine of the aircraft; I hate to think what would have happened if I had decided to raise the boom first.

It was about this time that the RAF Mk VI probe and drogue equipment was replaced with the Standard NATO Mk VIII probe and drogue. There was further training with the Vulcans and this was to lead to yet another trip to Australia. This time with three Vulcans of 101 Squadron who flew non-stop from Waddington to Perth in 18 hr 7 mins. Operational training was in full flow and the Squadron was to carry on training with the Vulcans, Sea Vixens, Victors and the Lightning aircraft, which would be the next aircraft to be tanked to the Far East. There was never a dull moment. One day we had a Valiant return with the hose and drogue still extended, some electrical fault had prevented it from being wound in, but the crew being professional landed safely.

Happy Days – 214 ground crew with a Valiant tanker – staging through Malta on the way to the Far East in 1961. Shaun Broaders is front row, third from left. (*Shaun Broaders*)

Tank change using Simons bomb hoist mounted on a four-ton Bedford Chassis. (*Shaun Broaders*)

There was also an incident when a Lightning lost its probe, you've guessed it, it was still in the drogue when the Valiant landed.

In December 1964 the Valiant aircraft were grounded because of severe metal fatigue in the main wing spars and were officially withdrawn from service at the end of January 1965, many of the squadron personnel being posted during the previous December and the rest during the early part of the new year. I consider myself to be one of the lucky ones as I was posted, in March 1965, to RAAF Butterworth, a unit I had visited regularly during our tanking exercises. Now my wife could see for herself the terrific hardship we suffered in the Far East!

214 Squadron disbanded on 28th February 1965, a sad day for all on the squadron and those associated with it. During the tanker years the squadron had a record that anyone would be proud of and history will recall that the first tanker squadron, 214 (FMS) Squadron played a significant role in the future of AAR. 214 Squadron was to reform at Marham on 21st July 1966 with Victor tankers. However, that is quite another story.

Chapter Fourteen

AN AEO'S STORY
Peter West

Peter West

In 1959 having completed a Signaller A course at RAF Hullavington I was retained there flying in the Varsity and spending most of the time sitting in the right hand seat as safety pilot. My CO, Sqn Ldr Stoten, had recommended me for a commission and whilst awaiting the selection procedure I was posted to Gaydon to train for the Valiant. This was not uncommon as there were already several airmen aircrew, mostly signallers and a few navigators, flying in Valiants. However, my captain during the OCU, Wg Cdr Jeffries DFC**, a FEAF veteran with a distinguished record, was not at all comfortable having an NCO on his crew. He was scheduled to take over command of 543 Squadron at Honington, the RAF's only strategic reconnaissance squadron and he made it clear to me that I would not be going there with him as it 'was, and must remain, an all commissioned unit'.

Meanwhile, whilst going through the ground school I was sent down to Biggin Hill for the officer selection board. This went well I felt and the board seemed greatly impressed that I had been selected for the V Force. The Valiant OCU was interesting and concentrated mainly upon the aircraft's electrical system. There was a brief period spent on the radio system, the STR18b, normally a reliable and efficient radio which, because of the aerial system, a supressed notch in the port wing root if my memory serves me correctly, was pretty useless most of the time – which was very frustrating. Little was talked of electronic warfare as the Valiant had only a limited fit and that was virtually useless. In effect my job was to be a wireless operator, much less demanding than my role in the Shackleton. Notwithstanding this I found flying in the Valiant exciting and very different from the piston-engined aircraft which had, up to then, been my workplace. My first experience of take-off was exhilarating and I watched, mesmerised, as the altimeter wound around at an alarming rate. When we levelled out at altitude the meter was still spinning, catching up with the aircraft. The Valiant was a well built, reliable and efficient medium bomber which operated well at high altitude. The pilot had, unlike the other V bombers, manual reversion; but this was very hard work and we rear-crew types found ourselves looking at our pilots' physiques much as we had done on the Shackleton, to see if they were strong enough to handle the controls safely. Gaydon was a pleasant location as we had Stratford-on-Avon close by, pleasant for evenings out, and Warwick within easy reach. However, my wife and children, my daughter had only recently been born, were still in married quarters at Hullavington so my priority was to drive home and be with them as often as possible.

When the course ended the senior AEO instructor told me that Wg Cdr Jeffries had kept calling him to see 'if Sgt West was all right and up to the job'. Poor chap was still unhappy about having NCO aircrew in his crew. Our co-pilot during this time was Barry Wood who would, many years later, become a close friend.

When the postings were announced I was destined for 214 Squadron at Marham, the RAF's only in-flight refuelling squadron. This sounded interesting and I looked forward to getting a crew. When I arrived at Marham it seemed remote and rather bleak. I was allocated a room in the sergeants' mess and on day one made my way up to the squadron clasping an arrival chit in my hand. Imagine my horror when I walked into the crew room to discover only officers there. I felt very self-conscious and alone. I needn't have worried, a smiling flight lieutenant pilot rose from his chair, asked my name and welcomed me to the squadron whilst introducing me to all those present. He was Eric Macey who was then a co-pilot but ended up as an air vice-marshal. I have always been grateful to him for his kindness to me, putting me at my ease and clearly realising my embarrassment. About 12 years later I met him at Scampton, he charged across the bar of the officers' mess and pumped my hand saying he remembered me from all those years before at Marham. He then said, "Well, aren't you going to congratulate me"? "For what"? I replied. "Last time we met I was only a flight lieutenant, now I'm a wing commander." "Ye Gods", I responded "I was a sergeant and now I am a squadron leader."

The OC of 214 Squadron at the time was Wg Cdr Mike Beetham, a quiet, unassuming but kindly soul with an enviable war record who was destined to become chief of the air staff and MRAF. I had only been on the squadron for a few weeks and had still not got a crew. I was on leave at my parents' home where my wife and children were staying, when the OC sent me a telegram saying I had been selected for a commission and was to report to the OCTU at RAF Jurby on the Isle of Man. Naturally excited I arranged for my wife and children to go to her parents' home in Scotland before clearing from Marham and heading for the Isle of Man. Unfortunately my course was going to cover the winter months so I would not see the island at its best. Of the one hundred cadets on my course only eight were former senior non-commissioned officers (SNCOs), all aircrew. We were christened 'Hairies' and tended to stick together. The directing staff left us alone and used us to educate the younger cadets in the ways of the RAF. One of the flight commanders, a flight lieutenant named Myers, introduced himself as a 'steely eyed killer of the skies', but blanched when he caught sight of former master pilot Laurie Gapper saying "Hello Mr Gapper. I didn't know you were on the course"! I asked Laurie if he knew Myers, to which he replied, "Know him, I taught the little bastard to fly!"

The course did little to prepare any of us for service as officers and at the end, whilst pleased and excited to be commissioned, I still had an awful lot of adjusting to do to fit in to the world of the officer. My posting was to 138 Squadron at Wittering. I had a week's leave with my little family in Scotland then drove down to my new unit, hoping to find accommodation for them as soon as possible. To my horror I was informed that there was no accommodation in the officers' mess but that I would be given a room in the sergeants' mess, not at all what I expected or wanted. I had, of course, to take all of my meals in the officers' mess and use the bar there. On day one I walked up to the squadron HQ, went into the crew room where, to my delight I met an old friend, George 'Ben' Benbow, who had been a sergeant signaller on my Shackleton squadron where we had been on the same crew. He was now a pilot and I admit that I clung to him like a leech. He and his wife were living in rented accommodation and offered to put me up, so I lodged with them until I found a place for my family. I was talking to Ben when the OC entered the crowded crew room, strode over to where we were standing and, ignoring me, started swearing at Ben about the strip of garden in front of the HQ. This, it transpired, was Ben's secondary duty and it was not up to the standard the 'boss' required. Then his eyes lighted on me "Who's this ****", he exploded.

Ben introduced me and when I said, "I'm your new AEO Sir", he retorted, "No you're******* not!" with his little piggy eyes boring into my soul. He then turned on his heel and stormed out of the room leaving me wishing the floor would open and swallow me up. My first day as an officer and gentleman.

The OC was Wg Cdr 'Tubby' Baker, a highly decorated World War Two bomber pilot, one of many 'Bomber Boys' recruited into the burgeoning V Force to get it going. Sometime later the adjutant came in and asked for Pilot Officer West. "The boss wants you." Frankly quaking I entered the OC's office. What a difference. He smiled, invited me to sit, offered me coffee and speaking gently without expletives explained that he had checked up and was told that I was to be the OC's AEO but this would not take place for several weeks, meanwhile he would ensure that I got plenty of flying. Whilst I was relieved at this news I still found it difficult to make the transition to my new status. However, I did not regret being commissioned and was delighted at the attitude of the others on the squadron who made me feel welcome and at home.

The Valiant, whilst a sound and capable aircraft, was not properly equipped as its electronic warfare fit comprised a very poor tail-warning radar code named Orange Putter, which was useless at detecting anything astern. The electrical system, which had to be monitored by the AEO, was very reliable and rarely caused problems. My crew, once the new OC arrived, were all very experienced but in some respects an odd bunch. The OC was Wg Cdr Henry Chinnery who came to the Valiant from his tour at Buckingham Palace as equerry to Prince Philip. The co-pilot was a delightful fellow named Ken Lovett who remained with us for about a year then got his own crew. The oddballs were the nav team. The plotter, who was indeed a first rate navigator, was James 'Jock' Copeland, a bachelor who had spent several years on Canberras. He tended to be overly fond of the bottle and, when drunk, could become violent, although for some reason he always allowed me to look after him. The oddest of the lot was the nav radar, George Hingley. George had been a pilot flying Wellingtons and earning a DFC. Post war he had eventually been posted to Canberras but his eyesight steadily deteriorated until he was regularly landing several feet too high. He was taken off flying but somehow persuaded the RAF to allow him to attend a nav course, at the end of which he was awarded a nav brevet and was posted to the Canberra as a nav. Then onto the Valiant as nav radar at which function he was frankly useless, his incompetence compounded by his being radar leader on the squadron – extraordinary. Flying with this crew was an unforgettable experience which I would not wish to repeat. I felt a bit useless and bored as an AEO on the Valiant, tried to help the nav as much as possible but was constantly frustrated by the nav radar's inability to do his job properly. Because Jock was an excellent plotter we won the navigation trophy during the 1961 bombing and navigation competition. Our Astro termination point error after a navigation leg of some 1,500 miles was only 0.7 miles. This threw into stark relief George's bombing errors during the same competition which were worse than the nav error, much worse!

George was certainly a character, fond of women, booze and gambling, but not necessarily in that order. On one memorable occasion as we were taking off on a night bombing sortie George switched on his 'mike', groaned into the intercom, then switched off. Later I asked him why he had groaned. "That's my TCIC groan", he replied. His response to my questioning the meaning of TCIC was, "Thank Christ I'm covered! If I drop any bad bombs I can tell the skipper that I was not feeling too well". Since he rarely if ever dropped good bombs his groan was in virtually constant use. His frequent lapses of acceptable behaviour were a source of continuous embarrassment to the squadron commander, an old Etonian, as was George's habit of addressing him as 'Henry'. I once asked George how he had managed to persuade the authorities to let him become a navigator. His response was that he had many good mates in the RAF whom he was happy to exploit whenever he felt the need for help. This, sadly, was all at the expense of both the RAF and Bomber Command. Thank goodness that all the other navigator radars on the squadron were efficient and professional in their approach to their duties.

An incident which is worth relating: prior to a Sunspot detachment to Malta, the boss wanted to check out the facilities at Luqa so he organised a pre-detachment trip for us to the island. He took one of the flight commanders with him, Sqn Ldr 'Polly' Perkins. Polly, a small, slightly built fellow, flew as first pilot with Henry in the right-hand seat. All seemed well at Luqa so after a few days we prepared to return to Wittering. Our take-off went smoothly enough but as we climbed the temperature in the cabin became hotter and hotter until the chinagraph pencil on my desk melted. All our attempts to rectify the problem were in vain and we were all at a loss to know what had gone wrong. Clearly the cabin pressure was rising inexorably. Later we discovered that Henry, not used to flying in the right-hand seat, had reached behind him on the right-hand coming and, having a tear in his cape-leather glove which had caught the increase/decrease switch, had pulled it to increase where it stayed. Polly then demonstrated the power of an adrenalin rush as he grabbed hold of the port DV panel handle and yanking with unexpected strength pulled the DV panel out of its position. This caused a rapid loss of pressure and we were, by this time, descending rapidly towards Luqa. In the meantime all of the co-axial cables behind the navigation crate had started melting and shorting out causing failure of most of our radar aids. Once back on the ground, and soaked in sweat, we had a busy and interesting couple of days replacing these cables and trying to get the equipment to work well enough to get us home. We made it home safely but with limited aids.

During the subsequent Sunspot detachment, which we all enjoyed, we were called to operational alert with the aircraft being loaded up with 1,000 pounders. This was caused by the troublesome situation in Iraq and our task was to bomb the airfield at Habbaniyah. This alert didn't last long and we were soon stood down as diplomacy triumphed.

Apart from Sunspot detachments our crew had no overseas trips as our captain said that having had a tour at the Palace he had had plenty of enjoyable overseas trips. Eventually he must have realised that we were less than pleased with this state of affairs so he loaned us to another pilot for a flight to Norway. Finally, just before the squadron disbanded, we flew, as a crew, to Southern Rhodesia, a most beautiful country where I was able to visit my aunt and uncle who had emigrated there after the war in which he had flown as a pilot. This trip made up for those we had missed out on over the preceding two years.

Early in 1962 we learned that the squadron was to be disbanded and all of the Valiants at Wittering replaced with Victor 2s. All but two of the aircrew were posted onto the Victor with myself and Flt Lt Dick Haven, a pilot, posted to 1 Group to fly in the Vulcan 2. I was delighted as at last I would be flying in an aircraft properly equipped with a full electronic warfare suite and my role as an AEO justified at last.

Chapter Fifteen

148 SQUADRON
Anthony Wright

Taff Foreman crew 148 Squadron. From left, Anthony Wright, nav radar; Ken Lewis, nav plotter; Taff Foreman, captain; Tony Gale, co-pilot and Daryl Pace, AEO. (*Anthony Wright*)

After completion of my navigation training I was lucky enough to secure a short holding posting flying Ansons on the Coastal Command Communications Flight at RAF Bovingdon. This was then followed by the Medium Bomber Force Course at the Bomber Command Bombing School (BCBS) RAF Lindholme. Finally, after another holding posting on the HQ 3 Group Communications Flight at RAF Mildenhall, again flying Ansons, I arrived at 232 OCU RAF Gaydon as part of 97 Valiant Course. It was there that I met my future nav plotter Sqn Ldr Ken Lewis. I was only twenty-one and, as he was about twice my age, I thought him nearly old enough to be my father. We both arrived on the course and were told that we were to replace a nav radar on 148 Squadron at RAF Marham. The difference was that Ken was going to replace him as flight commander but the shock that he got was that it was me who was going to be the replacement nav radar.

It wasn't long before Ken told me that although he was a navigator he was also a bomb aimer. In fact in 1957 on 49 Squadron, as the bomb aimer on Sqn Ldr Roberts's crew, he had dropped the second of the live H bombs named Orange Herald on Operation Grapple in the Pacific. The drop was in the Malden Island target area nearly 400 miles from their detachment operating base on Christmas Island.[12] With this pedigree I was very impressed to say the least. I also found out that he'd flown Lancasters during the war and later the American B-29 Superfortress, known to the RAF as the Washington, and wore, among other ribbons, the American equivalent of the DFC. However, being the man that he

---

[12] Chapter Eight Operation Grapple.

was, and now resigned to the fact that he was to be a nav plotter teamed up with a 'green' young nav radar, it is to his credit that in all our hours flying together he never once interfered with my crew duty.

At Gaydon with only us two knowing exactly where we were to be posted and in the knowledge that we were joining the rest of a ready-made crew at Marham, we teamed up with three others just for the course. Meanwhile, rank and 'who you know' often works and so I was immediately hauled off to the station commander's office to be introduced as the 'new lad' to a couple of Ken's old mates from Grapple on Christmas Island. Dave Roberts, his old captain, who was now the station commander and Arthur Steele, another captain who was on the third drop, was now the chief instructor. I had also met the then wing commander administration, Christmas Island, Dougie Bower. I just knew him as the station commander of RAF Lindholme during my BCBS course there and I remember his prophetic words which always haunted me; in one of our conversations, and knowing that I really wanted to be a navigator on Transport Command, he was the one who told me that the lines on charts and logs that I craved for were soon to be relegated to history and that despite my disappointment, radar navigation was here to stay.

Next I endured being thrown into the sea at RAF Mountbatten Combat Survival and Rescue centre in both single-seat and multi-seat life rafts. This was followed by a return to Gaydon to experience rapid decompression at 48,000 feet in a decompression chamber; later this procedure was undertaken at the Aviation Medicine Training Centre, RAF North Luffenham. Having completed the ground school and flying at 232 OCU, we departed for 148 Squadron to meet the rest of the crew.

RAF Marham was a large Bomber Command base with four Valiant squadrons. There were three bomber squadrons namely, 49 Squadron, 148 Squadron and 207 Squadron and 214 Squadron which was a tanker squadron. Defence was provided by the Bloodhound Mk 1 surface-to-air missiles (SAMs) of 242 Squadron with thirty-two launch pads. Finally, there was a USAF Custodian Detachment for the American nuclear weapons. These assets along with support equipment, engineering facilities, hangars, supplementary storage area (SSA) aka the 'bomb dump', buildings, married quarters and other accommodation used by station personnel made it a massive complex and viewed from the air on a Valiant H2S radar screen it appeared the same size as a large Norfolk town.

The squadron's crest was crossed battle-axes with the motto 'Trusty' often known as 'Thrusty' or 'Rusty' apparently dependent on what you thought of the squadron. 148 Squadron was also known to be the first V bomber squadron to drop conventional bombs in anger, in the Suez crisis. Wg Cdr Brian Burnett was OC the squadron during that time in 1956. I was to meet him much later in my career when he was commander-in-chief Far East. I served on his staff at HQ Far East Command in Singapore, a tri-service headquarters responsible for the Far East including Singapore, Malaysia, Hong Kong, Gan, the Malacca Straits and South China Sea.

As a 'new boy' on the squadron in 1963, I was aware of the many very senior aircrew, both in age and experience. Apart from Ken Lewis, there were at least seven other WWII veterans. Our tough Welsh squadron commander, Ken 'Dai' Rees, topped the bill. To win a bet he'd flown a Wellington bomber under the central span of the Menai Straits Bridge, been shot down in another Wellington over Norway and ended up as a POW in Stalag Luft III. There he was involved in The Great Escape as a tunneller. However, on the day of the break-out he was in the tunnel awaiting his turn to escape when the alarm was raised by the Germans and he had to beat a hasty retreat to his hut which almost certainly saved his life. Finally, he endured, and survived, the March of Death from Sagan. A truly remarkable man.

My first crew, along with Ken consisted of John 'Taff' Foreman as captain, qualified flying instructor (QFI) and instrument rating examiner (IRE), much later to be involved in the initial discovery of the Valiant metal fatigue problem whilst airborne from Gaydon. My co-pilot was Tony Gale and my AEO was Daryl Pace both of whom were older bachelors. It was a good crew to be on and one unusually with an AEO, rather than an AEOp, as out of the eleven crews on the squadron there were just three AEOs. The other crews had master aircrew and SNCO aircrew in that specific crew duty. These in turn were composed both of AEOps and signallers. In this respect the Valiant was unique being the only one of the three V bombers that was to carry non-officer aircrew. Later to redress any difference between a signaller and an AEOp the MOD, in its sometimes puzzling wisdom, decided that the RAF couldn't have signallers on Valiants with the 'S' brevet. To remedy this they sent our signaller aircrew off on a course and for them to return soon after, to continue their same job. However, much to some of the recipients' disgust this time they were wearing an 'AE' brevet. At this point it should also be noted that as well as AEOps and unknown to us at the time, and not on our squadron, there had also been the odd non-commissioned navigator serving on Valiants.

My first week on the squadron saw our crew, and other crews from our squadron, fly to RAF Scampton in white Valiants, along with other squadron crews from Marham, ready to stand in front of a long row of Valiants and to be inspected by General Lyman Lemnitzer Supreme Allied Commander Europe (SACEUR). It made the national newspapers which showed photographs of crews, each of which had to stand line abreast in front of their Valiant, while the general was driven down the line of aircraft in a highly polished open top Land Rover. He was effectively our overall commander as the three Valiant bomber squadrons at Marham were assigned as the Tactical Bomber Force to SACEUR. This meant that on QRA, and in the event of war, we had two American nuclear weapons loaded, side by side, in each Valiant bomb bay.

On return to Marham there was pressure to attain operational status as soon as possible. It was in everybody's interest on the squadron as another qualified crew meant another crew to take on the burden of QRA. In terms of the American weapons the first that the Valiant carried was the Mk 28 which was designed to be dropped from high level. However, by the time I arrived, in 1963 low level attacks were in practice as we had just switched to the Mk 43 which was termed a 'lay-down' weapon. The objective was to use these weapons against targets in SACEUR's plan. Of course that was if we went to war, alongside our US allies, against the Soviet Union. However, the UK also had a national plan whereupon if we decided to go to war without the Americans then we could do so with our own nuclear weapons against targets planned by us. In that case it would be with a single British nuclear weapon; initially the Valiant carried the Blue Danube

General Lemnitzer inspecting 148 Squadron. (*Glenn Sands Collection*)

which was then replaced by Red Beard. It was the latter that I was trained to carry at Marham and if that was used we were committed to a pop-up attack. For continuation training and standardisation we had to undertake bomber training requirements (BTRs). Much later when we changed to Strike Command (STC) they became STCTRs but as this was such a mouthful to say the old BTRs still stuck. However, this meant that crews joining much later erroneously thought that it meant basic training requirements. Whatever people called them it meant the same thing; for a start, hours of target study had to be undertaken each month, both on the SACEUR plan and national plan. Crews were constituted at this time and that meant individuals generally staying together on the same crew for at least a complete tour of two and a half years or even three years. This was for reasons of security on 'a need the know' basis, as a crew only knew their own targets and one other squadron crew's targets. Having knowledge of one other crew's targets was purely back-up cover in the case say of crew leave or sickness. To emphasise the point, members of a crew had to agree to go on leave at the same time. Not always a satisfactory way of operating when considering children, school holidays, working wives and time off in the summer which was wanted by most people. The captain of one crew was a member of the RAF Bobsleigh team and therefore they all used to end up having to take two weeks leave in the winter season. Obviously the rest of the crew, and undoubtedly their families, were not amused. Generally crews didn't always get their full allocation of leave. It was a time when leave was deemed to be a concession not a right.

As to the actual overall number of targets chosen by Bomber Command we were never told. This was something that was not questioned as anything to do with nuclear operations was on the same 'need to know basis'. What we did know, from our individual crew target study, was that there was clearly a large target list of both cities of significance and also opposing air force airfields whose aircraft posed a nuclear threat to the UK. As I recall, all my targets on Valiants both on the SACEUR plan and national plan were airfields often out in the middle of nowhere. In addition the distance that we would have to fly to strike them would almost certainly mean it was 'a one way ticket' for us. Lack of fuel for the return would have forced us to land back at some friendly country rather the UK.

I was six months into my tour when *The Great Escape* film was released in July. As our CO had taken part in the real Great Escape he and all the aircrew of 148 Squadron, along with wives and girlfriends, were invited to the release of the film at the Majestic cinema in King's Lynn. It was useful both for the cinema in a public relations exercise and for us as we got to see the film for free. We all arrived in our No 1 Home Dress, commonly known as Best Blue, uniforms which was part of the deal. We were also given the best seats and were placed on the front few rows of the balcony. After a few words from the cinema manager we all had to stand up to be seen by the rest of the audience. It was the nearest any of us would get to in achieving celebrity status.

US Mk 43 nuclear bomb.

Quick Reaction Alert (QRA)

On the subject of QRA while the concept was exactly the same throughout the V Force it depended on how the Group and the stations involved carried it out. The Vulcans came under HQ 1 Group, RAF Bawtry while the Valiants and Victors came under HQ 3 Group, RAF Mildenhall. The type of aircrew accommodation also varied from station to station, so too the length of duty on QRA carried out at one stretch and where the aircraft were parked and guarded. Each of the three Valiant bomber squadrons at Marham put one Valiant, along with squadron aircrew and ground crew, on QRA 15 minutes readiness called readiness 15. This was for 24 hours a day, 365 days a year. At Marham we lived in hutted accommodation, affectionately or otherwise, known as Butlins, on the other side of the airfield. Crews could use their time away from normal duties to plan future sorties, target study or train in the aircraft simulator. A crew vehicle was provided to ferry the crew around the station for this purpose and to enable them to get back to the QRA area within the 15 minutes or quicker if the need arose.

Of course we could also just relax in the QRA complex, play snooker or watch TV until we received a call out. One of the captains, a bachelor, who normally lived in the mess and who wanted to make his life more bearable, always brought his African Grey parrot along with him to share his room. Unfortunately, as I recall, the captain in question had to spend extra time cleaning his room before going off duty as his parrot wasn't too fussy regarding its own toilet arrangements.

The aircraft were situated just a few yards away parked in a circle on a ring of tarmac within an enclosed, wired and guarded area with huge wide gates. On a call out, the readiness state was upped from 15 minutes to 5 minutes to get airborne and thus readiness state '05' was declared. The gates to the aircraft compound would be opened; we'd then run to our individual aircraft, climb in and wait in the cockpit listening for orders via a telescramble lead connected to the aircraft direct to the bomber controller at Bomber Command Headquarters, RAF High Wycombe. A subsequent readiness state could then be upgraded, if necessary from the bomber controller, to readiness '02', start engines, taxi and then the inevitable scramble. The anomaly at Marham was that because we carried American weapons we also, whether we liked it or not, were accompanied in the crew compartment by a, normally burly, USAF service policeman. He was armed with a loaded pistol and had orders to shoot any of us if we attempted to start engines without the 'go ahead' from the president via SACEUR and the duty USAF custodian officer. The latter was always available, twenty-four hours a day, as he did his stint of standby, alongside us, within the QRA complex. However, if we did go to war with the American weapons I have reason to believe that the service policeman would have vacated the aircraft early voluntarily if we hadn't kicked him out!

This way of life could be carried on continuously for as much as up to two tours, as most aircrew, under the policy at that time, were locked in to around six years in the V Force, before moving on, possibly, with a posting elsewhere and then often still connected with the V Force.

Bomber Training Requirements

As already stated hours of war target study had to be undertaken each month. This and other training included sorties and emergencies in the aircraft simulator for pilots, simulator training for AEOs, weapon and navigation simulator training for the nav team, crew escape drills, life raft drills, lectures, crew checks in the air, pilot night flying currency, various approaches, landings and simulated emergencies to name but a few. These were all carried out over a six-month period. Annually, for a more realistic life raft drill, you had to jump into the sea from an RAF air sea rescue launch and then often winched up in a helicopter. For Marham this was generally in the cold North Sea. An aircrew standardisation and inspection team, commonly known as the 'Trappers', from HQ 3 Group put each aircrew member through their paces by means of oral tests on the ground and flying with them in the air. Finally, the Weapons Standardisation Team from the BCAS, later to be called the Royal Air Force Armament Support Unit, RAF Wittering examined all aircrew on their proficiency at handling nuclear weapons. This was both on a weapons simulator and out on the airfield at an aircraft loaded with a training nuclear weapon.

The biggest bugbear that crews had to endure was the crew classification scheme. Once operational, and ready for war, crews had to be classified in terms of their proficiency in bombing. You were expected to achieve combat status fairly quickly, then after a number of months having completed a successful BTR period combat star was attained, then select and finally select star. It wasn't until a few years later that they invented an even higher level called command. Thus after every six-month period it was possible for crews to move up the scale based purely on their bomb scores. However, these had to be achieved with the increasing difficulty of the reduction of yardage allowed around the target for each subsequent higher status. It can be seen that as the six-month BTR periods progressed, even if you were successful in achieving the next stage, it was often not far from the end of your tour that you attained select star, if at all.

As the NBS equipment was only accurate to 400 yards bombing from 40,000ft upping your status was not as easy as it appeared. The carrot to do well was that the higher you climbed the better the choice of a sortie abroad. This was known as a ranger. A western ranger was normally a sortie west to the United States while a lone ranger was to the east, generally Malta or Cyprus. It must be remembered that the crew classification rested on bomb scores and so the pressure was set fairly and squarely on the shoulders of the nav radar. This meant that for every classified bombing sortie that was carried out he was the only one really tested.

High Level Simulated Bombing

Simulated high level bombing in the UK was done by attacking selected targets not by dropping an actual bomb but by sending out a tone signal which ceased at the point from which the aircraft would have dropped its bomb. At high level the forward throw of the bomb, the distance that the bomb would travel in the air, could be five or six miles. On the ground a Radar Bomb Score Unit (RBSU) would track the aircraft as it approached the nominated target and plot where the bomb would have landed. It would then pass the score to the aircraft, in code. The result could be a direct hit on the target or a measurement in yards of the distance that the bomb was from the target.

In the early to mid 1960s there were a number RBSUs situated throughout the UK and were as follows: Glenrun (Scotland), Highfly then changed to Kenley (London), Ouston (Newcastle), Pinplot and Brantub (Norfolk and Suffolk) and Haydock (Merseyside). It was only when the Valiants were withdrawn from service that others such as Cider (Dunkeswell, Devon), Glasgow (Scotland), Spadeadam (Cumbria), Stornoway and Benbecula (Outer Hebrides), Tumby (Lincs), and West Freugh (Scotland) took their place.

The nav radar normally dropped the bombs and this was by means of his radar screen the name of which was H2S. As most targets didn't show or were cluttered by other radar returns other identifiable returns were mostly used as aids to the aiming point. These were known as target offsets and with the known measurement to the target north or south, east or west the nav radar could place his electronic markers, using a small joystick, known as the Control Unit 626, on the H2S screen over the offset and track the aircraft to the target. The two internal offsets potentiometers (pots) catered for radar responses up to 20,000 yards away from the target while the two external offsets pots extended to 40,000 yards. Moreover, if the pilot selected the steering to be controlled from the rear cabin then the nav radar could make corrections on his joystick and there was no need for the pilot to intervene at all and the nav radar could fly the aircraft himself to the target and drop the bomb. Certain targets and offsets still come flooding back today, Hallington Reservoir, Morpeth mental hospital, Flixborough Works, a hangar at Ouston Airfield, Goole, Haworth glass bulb factory, the Lion House in London Zoo, the reservoirs at Heathrow airport and even Sheffield town hall. However, the nav plotter also had a small commitment to drop bombs visually. This he did by lying prone in the visual bomb aimer's position situated underneath, and just aft, in the nose of the Valiant. The aiming was done through a T4 bombsight, the same system as used in the Second World War. To help him the nav radar would guide the aircraft to a certain point until the nav plotter was happy to take over. Whilst the nav radar didn't need to worry whether it was day or night on his H2S the nav plotter did have to be concerned at night. In fact one of the most popular targets at night for plotters was one of the London railway stations, either Euston or Liverpool Street, where because the lighting pattern around it was blue in colour and the target stood out from everything else in the area it was a gift to them visually.

The bombing runs weren't by any means all the same in execution or type. There was a host of different ones at high level. The standard straight and level run in was called a Type 2 attack, an 'S' weave towards the target to avoid SAMs a Type 2A, jamming from an RBSU or enemy was combated by a Type 2B, there was also a 2C and two different pop-up attacks were either a Type 2D or 2H. Finally, there were attacks practising with simulated malfunctions of our equipment, or indeed for real, called limited aids attacks. One interesting point was, and I have had agreement from other Valiant nav radars who like me then went onto Vulcans, that you could see things on a Valiant H2S screen, both high level and low level, that you couldn't on a Vulcan screen. All three V bombers had exactly the same H2S scanner and equipment and I've flown all three. The Valiant screen was the best for radar definition closely followed by the Victor and lastly, by a long way off, the Vulcan. Logically they should all have been the same. However, they weren't. Our only conclusion, which may or may not be correct, was that the scanner in each aircraft was contained in a different aircraft manufacturer's designed nose cone along with associated material and that made the difference.

At times to get in the vicinity of the target at high level, or over large areas of sea, where there were no recognisable features available, astro navigation was used and practised on a regular basis. The nav radar took the astronomical observations (astro), known as shots, on the sun, moon, planets or stars through a periscopic sextant while the nav plotter worked out the calculations and plotted the results on his chart. However, it was during the operation of the sextant that the Valiant showed, in my opinion, a lack of thought in design. The sextant was normally signed out from the operations staff and carried out to the aircraft by the nav radar. There it was inserted into a mounting in the roof of the aircraft but in a retracted configuration until later, when required, the nav radar would pull a lever on the mounting whereupon part of the sextant was pushed outside the aircraft without depressurising the cockpit. The problem with the Valiant was that the positioning of the sextant mounting was in the roof directly above the rear crew table in front of the nav plotter's position. Thus, it was the only V bomber that when using the sextant the nav radar had to climb onto the rear crew table to take shots trying desperately not to stand on the plotter's hands or chart.

Bombing Ranges used by V Force in the UK
Of course as well as simulated bombing we also, both nav radar and nav plotter, had an individual commitment to drop conventional 'iron' bombs. These were either high explosive (HE) or practice bombs. There were a number of Air Weapons Ranges (AWRs), as they were called, that we used both in the UK and abroad. In the late 1950s practice drops for Operation Buffalo (Maralinga) were carried out at, among other AWRs, Chesil Bank and Sandbanks, Dorset, while practice drops for Operation Grapple were carried out at Orfordness on the Suffolk coast. However, in the 1960s and later in the UK apart from one, namely Tain (Scotland), the rest were also around the coast. They were: Holbeach, Theddlethorpe, Wainfleet (all three in the Wash), Donna Nook (East Lincolnshire), Cowden (East Yorkshire), Pembrey (Wales), Jurby (Isle Of Man), Garvie Island (Scotland), West Freugh/ Luce Bay (Scotland) and Aberporth (a missile range off the Welsh coast).

The Valiant could carry a load of up to 21 x 1,000lb bombs, so too the Vulcan. However, the Victor could carry 35 x 1,000lb bombs. The maximum weight of a single conventional bomb that all three V bombers could carry was 1,000lb. These bombs were medium capacity (MC) or HE each with varying types of nose and tail fuses, either for ground burst (instantaneous detonation when hitting the ground or delayed detonation on the ground at times set by our armourers), or airburst (detonating in the air at a certain height). Dropping the bombs could be either singly or in a stick i.e. one after another from two to the complete load of twenty-one.

Before any bomb left the aircraft one unusual feature, that only the Valiant possessed, was a large door aligned fore and aft just behind the bomb doors and hinged such that it opened in the fore direction just stopping at an acute angle. It was called a deflector plate and opened slightly ahead of the bomb doors when they were selected open. The reason for this extra door was, we were told, designed to deflect the airflow from buffeting the bombs when the bomb doors were opened to facilitate a smooth departure from the bomb carriers. That was the theory. Apart from 1,000lb bombs, at times we also loaded up small practice bombs such as 100lb or 25lb which often looked rather out of place in such a large bomb bay. I always remember how precise the armourers were when loading, and then marrying, known as crutching, such bombs to the carriers. Clearly they had, by their loading schedule, a precise amount stipulated that, for the smaller bombs, they had to turn the knurled wheel on the end of the carrier to crutch the practice bomb rigid on the carrier. Unfortunately for them they hadn't been taught, as I had been by an excellent nav radar instructor at 232 OCU, RAF Gaydon, Bill Dobson. He told us all that he believed that the armourers always crutched the bombs too tightly against the carrier and this resulted in the odd hang-up from time to time i.e. despite the release pulse going through to the Electrical Mechanical Release Unit (EMRU) the bomb didn't drop off. He told us to uncrutch them enough so that you could waggle the bomb, by hand, from side to side. As the armourers were always at the aircraft to see the nav radar remove the safety pin or pins from the bomb and attach any necessary lanyards, they were horrified to witness me uncrutch the bombs. But remembering Bill's advice I ignored their remarks and gave them the reason. I never did have a hang-up dropping practice bombs whilst other crews did. Clearly Bill's advice was spot on. Bill also said that he always liked to drop clean bombs. I did too and used to wipe off any greasy marks left by the armourers with a handkerchief, much to the annoyance of my wife who had to launder them.

Life in the V Force was a merry go round of QRA, detachments both in the UK and abroad, exercises, regular working weekends for the whole station and flying for hours. The latter was the aim of, it seemed, every flight commander, squadron commander, station commander, HQ 3 Group and HQ Bomber Command. The Valiant sorties that we used to fly were mostly five and a half hours to six hours in length whereas later on flying Vulcans they tended to be around an hour or so shorter. It also transpired that whatever trip we did we had to endure one hour of 'circuit bashing' on return. Another wheeze by our hierarchy was that they considered flying directly from Marham to Luqa was too short a time in the air and in order to 'get the hours in' we often had to fly either up to Scotland and use an RBSU there or down to London with the same objective before heading off to Luqa.

At the end of each sortie there was the regular debrief with the squadron engineers whereupon each crew member sat opposite a crew chief, SNCO or JNCO relevant to that trade whether it be engines, airframes, armament, electrics, navigation instruments, safety equipment to name but a few. This was the time when we reported the serviceability, or any faults, encountered with the aircraft equipment. My opposite number specialised in the H2S and NBS. Therefore, I had to fill out a form inserting the hours the equipment was running and what time, if at all, the equipment failed and the possible explanation. Most of the time my equipment was serviceable and I duly signed the form handed it in on debrief when it then disappeared into the engineering electronics block nick-named 'The Gin Palace'. The normal procedure was that if anything needed attention it was rectified and to give feedback the form was returned to the user with appropriate remarks. The first time that I used the form I'd written 'serviceable throughout the sortie'. In due course I received it back from the OC Electronic Engineering Flight (EEF) and was amazed to find a row of Green Shield stamps attached to the bottom of the form. It was the vogue then to collect the stamps that were given by shops, stick them in a booklet, and exchange a large number for goods at a later date. Other nav radars were not surprised with the stamps episode as it was from OC EEF known to everyone as 'Woobs', a loveable bachelor who was well known for being more than slightly eccentric. His other antics of mention were making home brew in the officers' mess using the large airing cupboards near to his room for the fermentation of beer, giving us the enjoyment of the aroma wafting down the corridors. His room was also something to behold as he adorned the walls with sepia pictures of Queen Victoria, Lord Kitchener, past high ranking officers of the Royal Flying Corps and RAF and bare-breasted African tribal women. To top this he produced his own Christmas card each year. He put his camera on a tripod and dressed in a pith helmet, wearing a monocle and holding a glass tankard full of beer he thrust both his head and the one arm with the beer through a large empty picture frame and took a photograph of himself by remote control. This garb was his favourite because at one fancy dress ball in the mess he turned up in a jacket with tails, riding breeches and riding boots. He wore his pith helmet with a small union jack stuck on the top, monocle and carried a crop with a bunch of bananas hanging from a pocket. To complete the picture he carried a stuffed fox under his arm. On another occasion when we had an officers' mess casino night he spent the afternoon beforehand locked in his office, with a closed sign on the door. There were a number of his SNCOs inside teaching him poker.

Whilst he was the only one that was known to be eccentric there were others on our squadron who were clearly innovators. This was especially on the social side. One was an AEO who again lived in the mess and was into anything technical. I've seen him service and re-assemble a motorbike in his room with oil all over the place and once he proudly showed me the dashboard of his Austin Healey Sprite that he'd 'tarted up' with press to test (PTT) button lights wired in to check the systems. I was speechless at the time as the only place I knew that these PTTs normally appeared were on the AEO's position in a Valiant. There was also the time when he wanted a stereo system and so decided that he needed to build a cabinet for it. Swaffham market was a place that seemed to sell everything, including secondhand furniture, so off he went armed with a saw. The outcome was that he bought a piano on the spot, used the saw to remove all the highly polished valuable wood, left the remnants in situ and returned to Marham with the useful parts of his purchase.

Another incident concerns the meanness and the lengths that the civil service bureaucrats would go to thwart the aircrew. One of our more senior nav plotters

Bryan 'Monty' Montgomery who had previously done a tour on Canberras, and therefore had been around a bit, was not a person to mess with. For a co-pilot a pre-requisite for a captaincy was a successful completion of an Intermediate Co-Pilots Course (ICC). Thus at one point in his tour, if he were ready and had proved his competency in the right-hand seat, he would depart for the OCU to undertake the course. Once the ground school and simulator had been finished and the flying phase reached it was the co-pilot's squadron, and preferably his own crew, who would supply one or more of the rear crew to fly with him. On this particular occasion Monty was detailed to go and fly with his co-pilot John Gillingham on his ICC. As no MT could be provided to take Monty to Gaydon the RAF expected Monty to use his own car and get paid the normal, miserable, low rate of mileage allowance. However, there was provision under the rules that in certain circumstances a high rate nicknamed 'Golden Wheels' could be allowed.

With no MT available Monty asked the accounts department in station headquarters (SHQ) for the high rate and was told that it was not to be. Monty stood his ground and continued to demand that in that case the system should get him there. Undeterred SHQ then went to great lengths to scour the train time tables to try and find a train that went cross country and get to the OCU on time for the flight next day. This proved impossible in the timescale as they'd left it too late and with Monty still refusing to go SHQ, just to spite him, came up with a more expensive option consisting of an aircraft, fuel and aircrew time. It arranged for HQ 3 Group Bomber Communications Flight, the one that I'd served on, to provide him with an Anson to fly him to Gaydon.

By now everyone on the squadron knew of Monty's battle with SHQ and so that day we all watched from the crew room as Monty waved to us and walked out to the Anson, that had just come from Mildenhall, carrying his nav bag and flying kit. He got on board and the Anson took off. After some time we heard from ATC that the Anson was returning to Marham. Not long after it landed, out jumped Monty who then arrived in the crew room with a broad grin on his face. Apparently when the Anson arrived overhead Gaydon the cross wind component was too great for the aircraft to land and hence forced an early return. The next day a victorious Monty set forth for Gaydon, in his own car, having secured the 'Golden Wheels' rate of mileage.

One further episode involved our crew and showed how the civil servant in accounts hoarded the money. On my first Goose ranger we had to go to accounts to be issued with Canadian dollars, my first experience with the money, and as far as I was concerned dollars were dollars. However, on the first evening in the officers' club at Goose, as I was the junior on the crew, I was to buy the first round of drinks. When I handed over some of my dollars the barman looked at them closely and asked if I had any more to which we all replied in the affirmative. He then asked if he could exchange all of our dollars to which we agreed. It was only after the exchange had been done we realised that we had also been done. It turned out that the dollars, although still legal tender, were old and extremely rare and all fetched a great deal of money on top of their face value.

Life on Valiants was work hard and play hard. Hence, Friday and Saturday nights in the mess was the time to 'let your hair down'. It was quite normal to stay in the bar until closing time, then bribe the airman barman to stay on for an hour or two later. This was done by handing a tankard round the assembled throng who put money in it and thus remunerated the barman for his work out of duty hours. It was a win-win situation for everyone. When he went home and still more alcohol was required and he'd wisely turned the beer pumps off which, sometimes luckily, didn't always happen and we took full advantage, then phase two came into action. Woobs and the 148 Squadron captain, who owned the parrot, both had their tailor-made sports suits designed to incorporate an especially long inside pocket to accommodate a full size bottle of Johnnie Walker whisky. Needless to say, once the bar was closed, these bottles were magically produced to provide extra drinks all round. However, in the early hours of the morning we all started getting rather hungry and so at this point the normal procedure was to get in our cars and head off to one of our favourite pubs, The Feathers at Dersingham. This was a well-known watering hole situated on the outskirts of the Sandringham estate. There were always a lot of us after-hours drinkers but we always got a warm welcome from the host Jack and his wife whose sister, she regularly reminded us, managed the Penge synchronised dancing team. Where the heck Penge was, we really didn't know or care. However, because it was after hours there was a lock-in and, to be within the law, food had to be provided with a drink. They were a great couple and so always served us all, whatever the number, with egg and bacon. Why they went to such bother as a lock-in we didn't know as in the other bar looking across at us, looking at them, were the men with the big feet and macs, the detectives and police from nearby Sandringham!

Detachments and other Flying Exercises

Three months after joining the squadron I got my first western ranger in October 1963. The system was that you flew to Goose Bay, stayed the night, and then flew to Offutt AFB, Nebraska HQ SAC attacking high level simulated targets en route and having them scored by the USAF NIKE missiles sites in the same manner as our UK RBSUs. After landing at Offutt we were normally given a day off before returning to Goose the next day again attacking high level targets on the return trip. Another night was spent in Goose before returning to Marham. The standard western ranger, on average, lasted five days. This was before low level flying, both at Goose and Offutt, was also available to the V bombers.

The other regular detachment in the opposite direction on a lone ranger was RAF Luqa. Here the plan was to be based at Luqa but fly to North Africa and to the RAF El Adem bombing range just outside Tobruk at the eastern end of Libya and drop various practice or live bombs out in the desert. Later on when I flew Vulcans we did the same thing including Exercise Sunspot that were longer detachments at Luqa. However, other times we flew directly to El Adem on detachment from the

148 Squadron in 1963. (*Bryan Montgomery*)

UK and missed out Luqa entirely. El Adem had a tenuous link with Marham only because the latter was nicknamed 'El Adem with grass' and this seemed to stick as the stock phrase although blatantly untrue, both physically and climatically, to say the least. Some said that it was because both were situated in the middle of nowhere. On Valiants we always seemed to bomb during daylight hours on El Adem and normally dropped thirteen or sixteen 100lb practice bombs from high level singly with the nav radar using the H2S or sometimes with the nav plotter using the T4 bombsight. The locals were often a bit of a hazard, but only to themselves, as they liked to steal the lights that surrounded the target area that was lit up at night. Another of their ruses was to usher some poor old camel onto the range and hope that it would meet its end by one of our bombs. If the camel did expire then they would claim that it was a female camel, best of the herd, and due to give birth.

On one particular detachment to Luqa my nav plotter met an old mate of his, a flight commander nav plotter on a Vulcan squadron. Both had to do their BTRs of a number of bombing runs dropping weapons from their respective aircraft on El Adem range. However, it came to light that my opposite number, the nav radar on the Vulcan, couldn't remember how to use a circular 'slide rule' that we nav radars had to use to pass information to the nav plotters to set up their respective T4 bombsights. My nav plotter, flight commander, was ecstatic that he had one over on his friend as he had a nav radar that could do the business. So with great delight he offered my services free of charge, except for a beer or two, to instruct the other nav radar on the correct procedure. Thus a 3 Group aircrew on a Valiant got to instruct a 1 Group aircrew on a Vulcan. It couldn't have been better!

However, little did I know that the ending would have a twist for me later as, unknown to me, Vulcans would be my next posting.

Chapter Sixteen

LOW LEVEL DETERRENT
Anthony Wright

When the Government took the decision for the deterrent to go low level two things happened, the aircraft were camouflaged and we had to start flying low level. We normally bombed from between 40,000ft - 45,000ft and all the aircraft were painted white with an anti-flash finish. However, with a change of policy for the Valiants to attack at low level, this meant a change of livery from white to camouflage paintwork. During this transition period two stories, in which I was involved, immediately spring to mind. One saw all of us aircrew standing outside our hangar awaiting the landing of the first Valiant to be repainted in the new camouflage colours. We had been pre-warned by ATC of its approach and so aircrew from the other squadrons were also watching for its arrival. Rumour too was rife as word had been spread that the inventor of the camouflage had ensured that his initials were in the 'swirls' of the paint finish. On seeing the aircraft in the air and on landing and despite straining our eyes, moving our heads from side to side and trying, with great imagination, to decipher these so-called initials it was to no avail. It took some time for it to sink in that we'd all been had.

Of course once the aircraft had arrived, one camouflaged Valiant among the rest of the white Valiants, everyone wanted to fly it! Squadron servicing was still the system at that time and so until each squadron had at least one camouflage Valiant on its establishment it was difficult to get a sortie in one. However, when more new Valiants appeared for the next few weeks everyone wanted to fly in one; this turned out to be a bit of a disappointment because although the outside was pristine the inside was just the same. It hadn't seen one coat of fresh black paint. All the old scratch marks, scuffs and dents remained along with the unforgettable smell of cold metal, rubber, fuel and other common odours familiar to aircrew flying the aircraft. Needless to say as time went by, camouflaged aircraft were commonplace and white Valiants became rare. Therefore, it was quite easy to predict the next

The first camouflaged Valiant shown publicly in 1964 at an air display. (*Garry O'Keefe*)

outcome. When we got down to only one white Valiant on the squadron everyone clamoured to fly that one. So typical aircrew – always boys at heart.

Another incident concerning camouflage came to my attention many years later. It was March 2011 to be exact when I happened to read the Victor Association newsletter and Garry O'Keefe mentioned the first Valiant in camouflage to display to the public from an article in *Flight International* dated 28th May 1964.

'Making the first appearance of a camouflaged V bomber at a public display, a green and grey Valiant B(K)1 of 148 Squadron, RAF Marham, showed itself in the low level capacity at the annual RAFA Whit Monday air display at North Weald. Coming in very low and fast from the north, the Valiant proved how successful the new paint scheme is, for many eyes tried in vain to follow the commentator's injunction to spot "the Valiant approaching from your right".'

In fact I happened to be on the display sortie. It was on Monday, 18th May 1964 and lasted two hours and thirty minutes. We displayed at Church Fenton, Hucknall and North Weald. It was a single navigator sortie. Taff Foreman had left the crew a few months earlier and was posted to Gaydon. My new captain, Flt Lt Norman Bevis, was another QFI, IRE and a WWII veteran who had flown Lincolns, among other aircraft, after the war. He and my nav plotter didn't want to do a weekend trip and so I, as the crew nav radar, flew the sortie with another 148 Squadron captain, Flt Lt Phillips, and my nav plotter missing. The co-pilot and AEO I cannot recall. Clearly an historic occasion and I didn't even know that I had taken part.

I was also involved with the introduction of low flying. The powers that be decided that they would introduce low level flying training in the vast wastes of Canada. This was to take place from RCAF Goose Bay, Labrador. Two Valiants, both from Marham, one from 148 Squadron and one from 49 or 207 Squadron, were tasked to trial the first low level routes. Originally they wanted select or select star crews to be the guinea pigs. However, as is often the case it wasn't feasible because they also wanted a QFI pilot and a senior officer on the crew. My crew, although only combat star, ticked the two latter boxes and so we were selected to be the crew on 148 Squadron. Our next port of call, in April, was a visit to Joint Air Reconnaissance Intelligence Centre (JARIC) at RAF Brampton. There we were given a thorough briefing of what their predictions were of what I would see on the H2S, in terms of the radar returns and a description, and photographs of the target for the pilots just as we would going to war against the Soviet Union. Furthermore, our crew was also to be the first and only crew to trial the RAF's new prototype cold weather flying clothing. This consisted of a separate jacket and separate trousers. These could be worn instead of a flying suit or just the cold weather jacket on top of the flying suit, as my nav plotter on Vulcans often did in later years, even in Malta. In 1964 the safety equipment designers wanted to know how acceptable it was to work both in the new clothing and how it stood up in a cold climate. Our crew was to report back at the end of our detachment.

Low level flying over Canada. (*Anthony Wright*)

On 13th April two Valiants departed for RCAF Goose Bay to take part in Trial 478A. There were four routes to trial and our crew would fly two namely: Bomber Command BC/CAN 3 and BC/CAN 4 while the other crew flew BC/CAN 1 and BC/CAN 2. Our first sortie on the 15th was from Goose Bay to Hudson Bay and flying low level back to Goose Bay. The second sortie we completed was on the 17th and from Goose Bay to Val-d'Or and flying low level to Bagotville and then back to Goose Bay. We returned to the UK on the 18th and then later both crews went back to JARIC to be debriefed and put forward our assessment of the practicality of the four routes. Suffice to say that the radar returns predicted by their staff were extremely good and the routes, both crews agreed, were to be just what that was required. And so they were. I was still flying those original routes that I had trialled then, and other routes out of Goose Bay, many years later on Vulcans. However, it was not to be for our crew to be allowed to fly them immediately afterwards, despite trialling the routes. The reason quite simply was that the rules stated that you had to be at least a select crew classification to fly these new routes and we didn't qualify as we were still combat star. How bizarre!

Flying high level with an aircraft designed to favour the captain and co-pilot with ejector seats and the rear crew without was one thing. However, it was clearly going to be even a bigger problem to solve for the rear crew when needing to vacate the aircraft at low level. At Marham in 1963 the three rear crew seats occupied by the nav radar, nav plotter and AEO moved back and forth on rails like a car seat. However, later that year a team arrived to demonstrate the new rear crew seats. The aircrew from all squadrons were asked to assemble outside our hangars to witness the roadshow of three new rear crew seats, fixed to a platform, on wheels that would obviously be shown at the other stations with Valiants. The new three swivel seats were demonstrated and we all tried them out. The way that the seats worked was by moving a lever on the side of the assembly, which moved the back rest forward and one could swivel the seat to a position pointing towards the door/exit. Having done this and by pulling a small knob a gas bottle enabled the seat to inflate pushing you forward, at the same time removing a retaining pin from each of the two parts of the seat lap strap, which then dropped away, allowing you to vacate the aircraft, supposedly, more easily.

Flying and bombing at low level in the Valiant was certainly a change to the normal high level bombing attacks and an exciting experience especially for the pilots; however, not so exhilarating for those in the rear with limited visibility. Although the Valiant's elliptical windows on either side of the rear crew compartment did, to some extent, give a constricted view out each side. In the nav radar's case only, his window although the same size as the AEO's, had a surround such that the whole thing could be removed providing an exit, if need be, in an emergency or crash landing. We practised vacating via the normal entrance door, through the top with the canopy removed or through my window. The latter either head or feet first was quite normal for me in the emergency crew drill trainer which was situated in one of the hangars.

Another aspect to flying low level was the times when a lot of buffeting and a bumpy ride were experienced. Here the nav radar was affected most as it was a matter of him trying to synchronise his head bobbing up and down in time with the radar screen that was doing the same thing just to either get fixes for the nav plotter or to concentrate on a bombing run. Furthermore, the bombing runs weren't the long sedate run-ins that we were used to with a forward throw of the bomb from four to six miles. Here it was fast and furious with a forward throw of just over four hundred yards.

There were designated routes around the UK where low level flying could take place and along these routes were pre-planned fix points to aid navigation. In the early days of low flying all crew members were issued with a strip map which was a book of half million scale maps with the low level route drawn on it. These were originally produced by JARIC at RAF Brampton. The pilot's strip maps were heading orientated and smaller in size to the nav team who had strip maps oriented to the north. The maps were laminated so that we could mark with a chinagraph pencil. The nav radar's also contained small, square, acetate sheets with a prediction of the terrain radar returns printed on them, what the fix point or bombing fix point and target would look like on the H2S screen and the latitude and longitude of their respective positions. Matching the overlays to the H2S screen aided the nav radar in both identifying the navigation fix points so that he could update the equipment for the nav plotter or for use during the subsequent bombing run. The same procedure and aids were used for war targets. However, in that case a detailed large-scale map along with photographs from various sources were also added by JARIC. Not unlike high level attacks, the low level training attacks also varied and at times a practice bomb was dropped at the end of a sortie on a bombing range.

One innovation appeared in a few of the aircraft and was called a moving map display. This was inset in the table at the nav plotter's position under a transparent perspex screen. Before take-off the nav plotter loaded the parts of half million maps relevant to the sortie on cylindrical rollers. Once airborne the maps, that had been inserted and driven by the Ground Position Indicator 4 (GPI4), effectively rolled along track and showed the position of the aircraft in relation to the map. However, it was only as good as the maps that were loaded. If you had to divert somewhere for example and that map, or part of the map, hadn't been loaded on the ground then it was totally useless. Ken Lewis and I didn't have much confidence in the system. We weren't happy with its accuracy and to load it with selected maps was cumbersome so that we just didn't bother to use it at all. The equipment and its limitations were also confirmed by Air Commodore 'Spike' Milligan, then a flight lieutenant nav plotter whom I knew on 49 Squadron. The whole system was not at all accurate, and faded into disuse fairly quickly.

Spike recalls that he was also involved in trialling water-cooled suits on 49 Squadron. Although the air-ventilated suit (AVS) had been in service for some time, at least since 1956 on Canberras, and I personally went on wearing them flying Vulcans until they were withdrawn from operational service in 1984, I wasn't issued with one on Valiants. The AVS was a body suit made of fine fabric with a multitude of small pipes fixed all over it. Attached to this was a larger hose that could pass through the flying suit and be plugged into the aircraft supply at individual crew stations. Crew members could then, by turning a manual thermostat, either regulate hot or cold air to enter the suit to either warm or cool themselves dependent on the cabin temperature.

However, there must have been moves afoot to produce another form of individual cooling system as Spike trialled the water-cooled suit that, not unlike the AVS, was worn next to the body underneath the flying suit. It is believed that the water was fed into the suit from a reservoir situated somewhere under the rear crew table. The problem was that the suits leaked. Heating wasn't mentioned, so clearly no hot water was involved or indeed a kettle. Obviously the design did not come up to expectations and as I've not heard it mentioned anywhere else, the presumption must be that it was shelved.

There was no terrain following radar (TFR) in the Valiant. That came later in the Vulcan. Therefore, for low level it consisted of the nav plotter giving a running commentary on any obstacles, power lines, significant points that the pilots should see coming up, alterations of headings and time to the next turning point or target. Meanwhile, the nav radar would warn the pilots of the terrain ahead where he had 'cut off' on the radar screen which meant that the aircraft was too low for the H2S radar to see beyond, say a hill, and instruct pilots to climb. He also monitored and called out the aircraft's height on the Mk 6 altimeter situated alongside his nav panel. Throughout all this the AEO was contacting various airfields and agencies on the R/T giving position reports as we passed through their areas, receiving weather information and calling up the RBSU, if we were to drop a simulated bomb, or an AWR, if we were to drop a single practice bomb or stick of bombs. All in all it was a slick team effort by all members of the crew.

As to the cold weather clothing trials my nav plotter, as flight commander training, decided to send me down to Bomber Command Headquarters at RAF High Wycombe to give my opinion at a quite high powered meeting. He chose me out of the crew because the meeting was convened on a Friday and it was on the way to Hampshire. I went there every weekend to see my fiancée, later my wife, and it meant that I could claim mileage allowance just as far as High Wycombe and back. I duly said my piece to an audience, some of whom had never heard of the clothing and none had actually seen. I was amazed that I took centre stage and they listened intently to every word. My main point was that the clothing was made in an off-white, verging on beige, colour and while it might be satisfactory for escape and evasion in the snow it would be useless to attract attention for rescue. Therefore, I suggested that it should be dual coloured such that it could be turned inside out, if necessary, with day-glow colour on the reverse to aid rescue.

In the end the complete kit, jacket and trousers, came in just one colour. It was RAF drab blue exactly the same colour used for earlier flying suits. In later years the colour changed to NATO green but still not reversible. I was still wearing the same pattern 'Suit, Aircrew, Cold Weather Mark 3' in 1984. So much for my sensible input!

Bomber Command and HQ No 3 Group were continually initiating flying exercises of some sort or another. Names like Exercise Anchor, Exercise Kingpin, Exercise Hallmark and Exercise Co-op all conjured up specific sorties. They always included a navigation stage high level followed by a low level stage that involved bombing using either an RBSU or an RBSU followed by a practice bomb dropped on an AWR. Exercise Co-op consisted of flying over Norway, Sweden, Netherlands, Denmark, France and Germany while the more frequent Exercise Kingpins tended to involve flying over Norway and Denmark. Just a reminder here that the AEO who doesn't always get a mention was always involved in some way or other. On the larger assessed exercises when everyone was competing against each other in squadrons, stations and crews, the AEO was also assessed on the messages that he could receive and decipher via morse and his own dexterity with the morse or telegraph key that was fitted on the rear crew table at his crew position. The morse 'straight' key was the same type that my father had used, as a signaller, on Stirlings and Lancasters during the war, so that part had never changed. Although I do know one AEO who didn't think much of the speed that he could get out of the Valiant morse key and brought along his own sideways operating 'sideswiper' morse key and, as only AEO's can, fitted it up each time he flew to get the maximum words per minute.

One of the largest exercises, in aircraft number, that I took part in consisted of all serviceable Valiants, Vulcans and Victors in Bomber Command flying around a pre-determined low level route together with marginal time spacing, over East

Anglia. Norman Bevis, my captain, called me up to stand behind the pilot's seats to see this amazing spectacle. The planned route was so tightly confined in the area of East Anglia that it turned upon itself at various points. Therefore, as some aircraft were going one way others who were further back in the stream were going past in the opposite direction. How there was not a mid-air collision I'll never know. It was a disaster waiting to happen. The only thing going for it was that it was clear visibility all round. The planners must have learnt from it and the objections raised afterwards, because we never had another exercise like it again.

Exercise Mick, Kinsman, Mickey Finn and Dispersal Airfields

During the Cold War when the V Force was Britain's nuclear deterrent, the Valiants, Victors and Vulcans were based at nine RAF airfields namely: Coningsby, Cottesmore, Finningley, Gaydon, Honington, Scampton, Waddington, Wittering and Wyton. However, it was decided that it would be imprudent to have all the V Force assets confined to just the nine airfields. Therefore, in 1959 a list of thirty-six medium bomber dispersal airfields, including those already mentioned, was approved for ORPs and associated facilities to be constructed. Some airfields were to have ORPs to take four aircraft and others for two. These enabled V bombers to be parked on hard standings adjacent to the end of the main runway to ensure a rapid take-off or scramble. Thus, it meant that, in time of tension building up to war, V bombers could be dispersed to operate from other airfields. This was in addition to the QRA aircraft in the 1960s on 24-hour standby 365 days a year, when aircraft were already loaded with live nuclear weapons. Operating from their parent airfield, these would be the first used in the event of a sudden attack or declaration of war.

The other bases that were approved were namely: Aldergrove, Ballykelly, Boscombe Down, Brawdy, Bruntingthorpe, Burtonwood, Cranwell, Coltishall, Elvington, Filton, Kemble, Kinloss, Leconfield, Leeming, Leuchars, Llanbedr, Lossiemouth, Lyneham, Machrihanish, Manston, Middleton St George, Pershore, Prestwick, Stansted, St Mawgan, Valley, Wattisham, and Yeovilton. Later Bedford and Tarrant Rushton were added to the list although this was eventually reduced to just thirty airfields and amended from time to time.

Exercises whether generation of aircraft to fly or purely generation (i.e. the preparation of the aircraft) were commonplace. Exercise Mick was a regular one that was called for by HQ Bomber Command. All the station personnel were called in with the aim to generate as many serviceable aircraft, along with weapons, in order to simulate preparedness for war. Once the station achieved the required number of combat-ready aircraft the Mick was terminated. A favourite time to call it was late on a Friday afternoon, just as most personnel were going away for the weekend or had already left the station. Another time at Marham a Mick was called on a Sunday but our squadron commander wanted to go one step further and open up our squadron's facilities at our dispersal airfield at RAF Manston, Kent. This was just to check that everything was in working order. Our crew was selected as the operations crew and told to drive down and activate our mini operations room there. As we couldn't contact our co-pilot, Tony Gale, who had gone to London for the weekend, we drove down leaving the squadron, who were generating aircraft, to continue to try to contact him. After activating the dispersal and finding that everything worked, some hours later, it was around 6pm, the exercise was terminated and we were recalled. At this point Tony arrived. He'd eventually made it. As he was getting out of the car he received no sympathy, but was just told to get back in his car as we had finished and it was time to drive back to Norfolk. Tony didn't say a word.

Apart from this one-off incidence, with a Mick being added to by our squadron commander, the dispersal airfields were activated regularly by aircraft flying to them on exercises called either Kinsman or Mickey Finn. The former exercise was a squadron dispersal of one or more aircraft while the latter involved a number of squadrons not only from one station but, at times, also different aircraft from different stations. Squadrons all knew their designated dispersal airfields and crews could stay there for two or three days before being scrambled on an exercise bombing mission. As already mentioned one of the dispersal airfields for 148 Squadron was RAF Manston. At this point it should be noted that whether it be for real or for exercise, the standard operating procedures for the handling of all nuclear weapons were exactly the same. Therefore, weapons live or training were given the same security on convoys from the supplementary storage area, within or without the confines of the station to the aircraft to be loaded. The aircraft were then guarded by armed RAF policemen and their dogs. To an onlooker there was no difference. As there were not enough training weapons available throughout the force, live weapons were sometimes loaded to complete the loading schedule. However, the latter were then downloaded and returned to the bomb dump leaving only those aircraft loaded with training weapons, and those without any weapon, to get airborne.

While the engineers, armourers and aircrew were in the process of preparing the aircraft to fly to the dispersals others were occupied too. Ground support in the form of engineers, MT, armourers, aircrew operations staff, caterers, RAF police and their dogs to name just a few of the trades, along with equipment, would already have been despatched by road, and sometimes air, to the designated airfields. On arrival the support party would open up the detachment facilities, sometimes near to the ORPs and mostly situated on the other side of such airfields away from the main operating areas, ready to receive the incoming squadron aircraft and aircrew.

In due course, back at the home station, the order to disperse would be given and aircraft would fly off to the dispersal airfields. On landing they would then be taxied to the ORP, parked, an after-flight service carried out, refuelled and if everything was serviceable the aircraft and crews would be ready to scramble in minutes when ordered by HQ Bomber Command.

Accommodation did vary from dispersal airfield to dispersal airfield. While some aircrew were accommodated in an officers' mess, others, including myself at times, experienced a rather different way of living. One such form of accommodation was that each crew was allocated its own five-man caravan. In very basic terms it consisted of a long, rectangular, green metal box (see photo in plate section). Effectively it was a smaller version of a shipping container that was divided into five individual compartments each with its own outward opening door. Each door had a toughened glass window that had a hinged metal flap on the outside that could be closed for complete privacy.

Each compartment had a narrow caravan bunk that was about six foot long and stretched from one side wall of the compartment to the wall on the other side. Any aircrew over six foot found it extremely uncomfortable. Above your head were some small cupboards but nowhere near enough storage for all our flying kit and nav bags. If you swung your legs out of the bunk you could sit looking at a tiny metal washbasin that was covered by a hinged wooden flap set at an angle to be used as a writing desk or raised when you wanted to wash. Under the wash basin, and next to the floor, was an unprotected long and deep metal bar heater that got almost red hot when switched on and almost suffocated you with the heat in the middle of the night with the door closed. Not to suffer serious burns by accidentally touching it in such a confined space was a miracle. I believe that there were only two settings, on or off. However, to refrain from using the heater, what with the extreme cold outside, it would have been freezing inside. With the longer side of the compartment being no more than six foot, the measurement of the shorter side was just over three feet. The whole atmosphere was claustrophobic, especially closed in at night. Water, only enough to wash one's face, was provided from a small water tank situated in the roof of an end compartment along with the main fuse boxes for all five compartments. The caravan itself was connected to mains electricity outside.

Years later I did see our aircrew caravans being used but for quite a different purpose. Recycling had clearly started early in the RAF. At RAF Wittering the compartments were being used by 5131 Bomb Disposal Squadron to store their bomb disposal equipment while at another RAF station I espied the RAF police dogs being housed in them – I did feel rather sorry for our canine friends.

The planning room, small operations room, ablutions area and varying messing arrangements were generally all nearby, often in small buildings, some of which were huts. In the freezing cold of winter on a windswept airfield, often muddy underfoot, neither this nor roasting inside the caravan were enjoyable experiences.

After two or three days, the time span constrained purely for exercise purposes, the order was given for all aircraft to get airborne on a simulated war sortie. With our ground support watching, a scramble take-off from the ORP saw four aircraft take to the skies in quick succession. The task of the personnel we left behind was also now completed and while we were flying they were clearing the site, packing up equipment and closing down the dispersal facilities until next time whether it be for an exercise or for war.

The first Exercise Kinsman that I took part in was just one aircraft with our crew who dispersed to Tarrant Rushton, Dorset. We flew the Valiant down to the airfield while a small support party, all coming from Marham, consisting of Flying Officer Dick Ives, from ATC, the senior medical officer, why the latter I'll never know, and an RAF policeman and his dog. They all drove down by MT and private cars. The accommodation was much better than we had anticipated as we were booked in at The Red Lion in Wareham. Not long after we all met up in the bar the senior medical officer, who liked a beer or two, pranged his brand new car in the car park. So we had to commiserate with him over the odd tipple or two to cheer him up. We were also joined for dinner by my fiancée who had driven all the way from Hampshire through the New Forest in her BMW Isetta Bubble Car. After a decidedly pleasant evening and one night away we flew back to Marham the next day.

Chapter Seventeen

VALIANT TALES

Listening to the tales of the aircrew it is quite clear that what they all looked forward to was a trip overseas, away from the demanding life of the QRA. Another theme, not often discussed, is how the bodily functions were managed on a long flight; in my Victor and Vulcan books the subject did not surface but as this is the final V Force Boys book the problem is highlighted. However, what is not said by the story tellers is the rather poor standard of the equipment provided.

Russ Rumbol was on Lincoln bombers and then started the NBS course in 1955. Before joining the OCU at Gaydon as a nav radar, he was on loan to Vickers Armstrong, test flying Valiants. He joined 90 Squadron at Honington in 1957 and left in 1961.

A western ranger was one of the delights of the V Force. We flew from our UK base to Goose Bay in frozen Labrador. After a short stay there, which was often fun in itself, the next stop was Offutt, no less than the headquarters of SAC.

Our first western ranger was in March 1960. Labrador was bitterly cold. The rest of the crew were sensibly in the warm mess. I was at the aircraft. The purpose of these rangers, apart from a nice jolly, was to make the crews self-dependent without the backing of the main base. We sometimes carried a crew chief but we were expected to look after our own equipment. Mine was a device called NBS, our main navigation and bombing tool – vital on the real thing. On the leg to Goose, mine had a small fault which could have been ignored – but not by me. Full of unjustified self-confidence, I set about fixing it, ignoring the extreme cold. In fact, I made it worse and, on the leg to Offutt, I wasn't much help to my brilliant co-navigator, Ken Hunt.

Offutt was a typical American base, except that it was the headquarters of SAC, the epicentre of the Western deterrent. We were shown around but it was a public tour – not just because we were in the V Force. I was surprised by how much was revealed; I'd have thought it was mostly classified. No one could have failed to be impressed by the efficacy of the system. It might have done some good for our doubters back in the UK to see it.

Then we hit the town. Omaha, Nebraska, is about the size of Nottingham and I can remember little of the town itself except that it appears to be deficient of public toilets. That's how the fun started. It was suggested to us to make use of a bar.

Inside, it was very obviously St Patrick's Day. Everything was green, including the beer. How horrible, we thought. Then a green pint appeared in front of each of us. Before we could protest, the barman said, "Compliments of the gentleman down the bar". Then advanced a small, very tough looking man, the image of

James Cagney, with his hand outstretched: "The name's O'Brian."

Oh dear, here's trouble! We knew that many Irish Americans hated the English and, when drinking on St Patrick's Day, in particular. Luckily there were four of us. But when we returned his greeting, he exclaimed, "Say, are you guys English?" He turned out to be a lovely chap and incredibly pro-British. He had himself been in the US Navy in the war. His father was in the IRA in the early years of the century and left for the United States when things got to be too hot. In the First World War, he didn't wait for 1916 but joined the British army, got the MM and died fighting for the old enemy at Mons. His son had a similar affection and seemed to delight in our company, pressed more green beer on to us and gave us each a silver dollar. Mine is now in the custody of my grandson.

Food was clearly called for now but it was not to be. The next place we made for was like the previous one but more so. I don't remember sitting down to a meal but I do remember a great deal of noise and a splendid welcome. If it were not for all the green, you'd have thought it was St George's Day. We weren't able to put our hands in our pockets once.

It was election year and a fellow called Kennedy was on everyone's lips. I found this surprising for I knew that Omaha was in the mid-west (trust me, I'm a navigator!), which I thought was Republican territory. But Kennedy it was, as all the costumes and beer mats said. Just two years later, we had the Cuban crisis and disaster was saved by the statesmanship of Khrushchev, Macmillan and Kennedy – not to mention the Anglo-American deterrent. Put some other names in there and who knows what would have happened.

We told tales, we sang songs and told insulting transatlantic jokes, taken in good humour on both sides. "The Yanks are flying Fortresses at forty thousand feet but they only drop a teeny weeny bomb". "Yeah, you Brits will always fight to the last – to the last American." "Our British beer wasn't warm in the war. It was cold when we made it – you took three years to get here."

Our jokes were down to earth but not too rude because there were a good many women in the bar. A very attractive one took quite a shine to Brian Warwick, our tall, blond co-pilot. She likened him to Tab Hunter. At that time of course, Mr Hunter had not 'come out', otherwise Brian would have thought it a doubtful compliment. I seemed to gravitate to the mature ones – just my luck for I was only twenty-seven. To one lady, I gave the benefit of my vast knowledge of American politics, from the Monroe Doctrine to the Marshall Plan. She seemed impressed and was very attentive and polite. When we left, she gave me her card (which I still have). It said "Mimi Olsen, Democratic Chairwoman for the State of Nebraska. 'What a berk,' she no doubt thought.

Came leaving time and we were full of almost tearful regrets. "Come and see us again the next time you're here." We were in high spirits and full of bonhomie but not drunk – it was only American beer after all. In the taxi back to camp, reality sunk in. At 0800 the following morning, we were going to fly back. How was it possible? At Offutt, we were met by the crew chief. Our aircraft was u/s and would take at least a day to fix. We almost kissed him. The next day, we resolved to be sensible. We would not take up all the offers we had but have a nice quiet time. Maybe we'd go to that bar again and have one or two drinks at the most. At the bar, we saw the attractive lady who thought she'd said goodbye to 'Tab Hunter'. Seeing us and especially Brian, she exclaimed "Well, blow me down!" This expression I thought was limited to Popeye!

> **Bryan 'Monty' Montgomery** graduated from Leeds University in 1956 with a degree in French. He was called up for national service, decided to stay in the RAF, took a short-service commission and was trained as a navigator at Thorney Island. He was granted a General List Permanent Commission shortly after. After converting to Canberras at Bassingbourn he was first posted to 249 Squadron in Cyprus before being posted as a plotter to 148 Squadron at Marham, where we became part of SACEUR and went into the low level role. He was grounded at the same time as the Valiant in 1964 to use his French and then was taught Russian so in 1971, with his linguistic ability limiting his RAF career he exercised his option for retirement and left the RAF.

Our crew on 148 Squadron consisted of Flt Lt Bob Fox captain, Fg Off Pete 'Kiwi' Dwyer as co-pilot, myself as nav plotter, Flt Lt Bob Downing as nav radar and Master Aircrew Jerry Bailey as AEOp.

Some of the more unusual aspects of life on our crew included a trip when Kiwi came out of the co-pilot's seat to use the toilet down the back near the rear crew. While he was there I noticed that the cabin pressure was rising and we were at 4,000ft and increasing pressure. Bob asked me to reduce pressure with the pressure release handle on my table. What no-one at 232 OCU, Gaydon had told us was that the pressure came off in the first 20 degree of travel or so. I turned the lever through about 45 degrees and all hell let loose. We immediately de-pressurised, the cabin filled with mist, papers were everywhere and we discovered later that I had jettisoned the 'dust-bin lid' somewhere over Scotland – it never was found, as far as we knew! With the cabin pressure now at 40,000ft plus and Kiwi not on oxygen he collapsed on to the floor. Bob Downing got out of his seat to attach Kiwi to a trailing oxygen hose supply and Bob Fox initiated an emergency descent. I then noticed that he was exceeding the aircraft limit and he pulled the nose up sharply. When all the excitement was over we recovered to base at 10,000ft. On taxying into dispersal the crew chief came on line and asked what the heck we had done to his aircraft, but in rather stronger terms. He even claimed that we had bent the pitot tubes, but I think that was an exaggeration. After the subsequent unit inquiry we lost Kiwi and got a new co-pilot, and we re-classified from select star to combat.

Western rangers to Offutt were always very popular, but ours were always odd. In May 1963 we set off for Goose Bay, but halfway across the Atlantic I stood up at my desk to get something and as I sat down I caught the release handle on my parachute – a new handle which was being trialled. It was a metal lever on the front of the pack, hinged at the top and operated by pulling the bottom of the lever upward. The pilot chute attempted to deploy, but I quickly sat down on it and remained sitting on it until we reached Goose, where the safety equipment section re-packed it and told me to be more careful. However, we never made it to Offutt, because on landing Bob Fox had got the nose too high and took the tail skid off. A new one was sent to Goose and we were told to come straight home.

For our second western ranger we landed at Goose with a temperature at -22°C. It was strange taxying to dispersal through purple banks of snow – the Canadians spray the snow to make taxying easier. As we came to dispersal the hangar doors opened and we went straight into a heated hangar – very pleasant.

The next day we set off for Offutt, expecting to do a six-hour trip via various RBSU sites. On calling up the first RBSU, we were told that no ranges were open that day – it was George Washington's birthday and no-one was working. So three hours later we landed at Offutt after a very short day. After a day off, we made our way back, uneventfully, to Marham.

In September 1964, just after the Valiant had started with cracking spar problems and aircraft were being grounded, I was asked to stand in for Les Wheeler, nav plotter in Ken Marwood's crew, to do a series of appearances at various airfields and shows for Battle of Britain weekend. We did all the normal stuff, tight turns, fast runs, steep climbs and so on. After we landed back at Marham, word came that one man watching our display at RAF Waddington had had a heart attack and died. Someone suggested that it was a Vickers stress engineer. This was never resolved.

Our crew were also among the last to fly the Valiant we think. We did a two-hour night sortie on 7th December, and the whole fleet was grounded on 9th December 1964.

Brian Loveday joined 207 Squadron at Marham as a co-pilot in February 1964 as his first operational tour after joining the RAF straight from school, and he only had less than a year on the squadron before moving on to the Vulcan.

It's such a long time ago and memories are somewhat faint. I don't seem to have photos but one story from my early days as a very green, young pilot on the squadron remains in my memory. I was teamed up with Flt Lt MacLachlan (captain), Flt Lt Jackson (nav plotter), Fg Off Phil Hawken (nav radar) and Sgt Elliott (AEOp).It was fairly obvious that there was some acrimony, for whatever reason, between the captain and the AEOp. While flying on a long high level night sortie the captain said to me that his right foot was becoming 'warm and wet' and asked me to take control. Having taken control of the aircraft I watched with some bemusement while the captain took off his sock and wrapped it around

the back of his ejection seat into the face
of the poor nav radar who was far from
impressed to being hit in his face by a wet
sock. At this point the AEOp was creased
with laughter. Unbeknown to me the nav
radar had got out of his seat to relieve
himself into the 'pee bottle', which was
mounted on the back of the captain's seat.
Only the AEOp knew that tube into the
pee bottle had been disconnected and been
redirected under the captain's ejection seat
and into the captain's right flying boot.
After a few harsh words post flight I don't
think that the two ever flew together again!

Co-author Anthony Wright holding
a pee tube.

Eric Macey struggled through basic
FTS but a respectable advanced FTS
placing earned a Hunter Conversion
Course, though Duncan Sandy's axe
curtailed his days as a fighter pilot
with 263 and 1 Squadrons. In October 1958 he joined the rapidly-
expanding V Force, initially flying as a Valiant co-pilot on 214 Squadron
then engaged on in-flight refuelling trials. This experience resulted in a
Vulcan captaincy on 101 Squadron (then training for a non-stop flight
to Australia) on which, over the next several years, he served as training
officer and squadron commander (and which formed part of his wing
when he was OC RAF Waddington).

I joined 214 (Valiant) Squadron as a flying officer co-pilot on 18th February 1959
being crewed initially with the squadron commander, (then) Wing Commander
Michael Beetham. Following his posting in June 1960, I flew with the new
squadron commander, (then) Wing Commander Peter Hill and remained in his
crew until posted to a Vulcan captaincy on 5th December 1961.

My first story recalls a Trial 306 Tankex as co-pilot to Squadron Leader John
Wynne. In early April 1959 we were the primary of two 'tankers' positioned at
Luqa, the other being captained by Squadron Leader John Slessor, with the task of
refuelling a third squadron aircraft attempting to fly non-stop to Kano in Nigeria.
On the morning in question, the wind was blowing strongly from the north-west so
the duty runway was the short Runway 32 beyond which was a deep quarry. Since
a safe take-off within the distance available was doubtful, I suggested to John
Wynne that we use the exceptional 200 rpm 'overspeed' (allowing a maximum of
8,200 rpm rather than the usual 8,000), together with the available water methanol
injection. John Wynne concurred with the recommendation to use water meth

but thought 8,200 rpm unnecessary. We laboured down the runway and as we clawed into the air just a few feet short of the end of the runway (and the lip of the quarry!) John Wynne pressed the transmit button and instructed, laconically: "John (Slessor), you'll need eighty-two."

My next story relates to the record-breaking non-stop flight from Marham to Changi on 25th/26th May 1960 when I was co-pilot to Squadron Leader John Garstin. Before take-off we had a side bet, the loser being the first to use the pee tube. Take-off from Marham was to be followed by three mid-air refuellings from tankers based at Akrotiri, Karachi and Gan. Following the first night refuelling over Turkey, John handed over control and had a snooze – during which time I had a surreptitious pee. John awoke in time to complete the refuelling over the Indian Ocean, following which he was clearly 'cross-legged'. He eventually, and reluctantly, resorted to the relief tube and conceded 'defeat'; and only then did I admit to my earlier use of the tube!

My final recollection is of completing two trips in the Marham flight simulator (on 9th November 1959 and 12th December 1959) as co-pilot to the legendary World War II veteran, Group Captain Leonard Trent VC, then OC RAF Wittering. As station commander, Trent had no co-pilot of his own whilst his station had no simulator so he was obliged to 'borrow' a co-pilot and to visit Marham to complete his obligatory training. On one 'flight', the instructor (Flight Lieutenant David Bridger) initiated an engine fire and failure on take-off, when Group Captain Trent's immediate reaction and instruction to me was: "Feather number 3".What nostalgia!

> **John Foster** had thirty years' experience and involvement in the Cold
> War starting on Shackletons in January 1959 with 42 MR Squadron at
> St Mawgan as a navigation instrument fitter. He was commissioned as
> a navigator in 1961 and went on to serve as a navigator (radar) on the
> operational squadrons of all three V bomber types as a specialist on
> weapon systems and flight refuelling. Here he tells of his experience as
> nav radar 214 Squadron Marham RAF. In 2012 the BBC asked him to do
> four radio broadcasts for their Cold War archives in the form of scripted
> interviews.

On completion of the Valiant OU in the first quarter of 1964 I was given what was considered a very plumb posting to 214 Squadron at Marham together with another first tourist Dave Ellis. This was a change in policy because until then only experienced aircrew served on tanker squadrons. The reason for this was that although the tankers in theory retained their bombing role they were actually employed solely on in-flight refuelling. The extra training required for both roles and the high demand for flight refuelling at the time meant that the squadron had to be entirely excused from QRA. There were four Valiant squadrons at Marham and four NATO targets had to be covered. Unfortunately, the three main bomber squadrons had to cover the QRA commitment for 214 Squadron as well.

Flight refuelling in those days was extraordinarily primitive. The Valiant used the 'hose and drogue' system pioneered by Alan Cobham. It had just one system fitted, a HDU mounted in the bomb bay. The fighter or receiving aircraft's probe fitted into a 'basket' at the end of the hose. Later baskets were made from fabric, but the Valiant had an all-metal basket. This meant that the receiving aircraft on making contact had to be extraordinarily precise to avoid causing damage to either aircrafts' refuelling equipment. The early marks of Lightning were a particular problem. Although much loved by the pilots and the public at air shows, they carried so little fuel that the tanker aircrew jokingly referred to them as 'fuel emergencies from take-off'. A further significant difficulty was that the Lightning's probe was awkwardly positioned to the side of the cockpit.

I remember a particular incident while tanking a Lightning over the Mediterranean on route from Malta to El Adem in Libya. The Valiant cruising speed was far too slow for the Lightning, which had to fly well below its drag curve. On this occasion after fuelling the pilot vented his frustration by giving us an impromptu air display. This was quite a spectacle, but he quickly burnt up his meagre fuel load and had immediately to take on more fuel. Unfortunately, the visibility was poor and he became disoriented and lost contact with the tanker. My captain passed him the tanker's heading so that at least the two aircraft would be travelling in the same direction. The Valiants' H2S radar like all V bombers was mounted in the nose of the aircraft. Therefore there was a dead space behind the aircraft. The Lightning was very close to ejecting when I managed to find him by using the special 'fish pool' facility and steer him towards the tanker. It was with considerable relief all round when the Lightning pilot made contact first time and I was able to broadcast the magic words 'Green Light, Fuel Flowing'. Failure would have meant a certain ejection in an area with no known search and rescue service. The drinks were certainly on him in the bar at El Adem. Fortunately, the RAF was also equipped with Javelins and these formidable war planes had an excellent range. Refuelling Javelins far from diversion airfields was far more relaxed.

Our OC was a Wing Commander Ken Smith. He was openly and affectionately known by all as Kipper Smith. I asked why and the reply was that he, like a kipper, was two faced, no guts, yellow and stank. It was said with a smile because everyone knew that no man deserved the nickname less than him.

Sadly, my tour came to an abrupt end on 9th December 1964 when the entire fleet of around 107 Valiants were grounded and subsequently scrapped. The squadron had a brilliant end of era party and concert when the entire squadron fund was blown. I went on to serve on Vulcans and Victors as well as becoming a NATO staff officer at Rheindahlen in Germany during the Cold War. On the early demise of the Scampton Vulcans I had a surprise posting and became the first navigator to control the Ballistic Missile Early Warning Centre at Fylingdales

Yellow Sun H bomb.

on behalf of the C-in-C North American Defense and C-in-C Strike Command. In this role I was personally responsible for issuing the 'four minute' warning of a ballistic missile attack (fifteen minutes for the USA and Canada). I ended my Cold War involvement by commanding the RAF's nuclear weapon training squadron at Wittering. While at Wittering I created a museum of obsolete nuclear weapon casings to serve as a brief introduction to the many handling and familiarisation courses. Examples found rotting on army training areas included Blue Steel, Yellow Sun, WE177 and Red Beard. The weapons were refurbished and painted with authentic markings by volunteers at Wittering. On trips today with my grandsons to various air museums I am delighted to discover that many of these weapons are now on public display.

Last but definitely not least is **Keith Walker**, Dick Hayward's co-pilot mentioned in Chapter One.

I was posted to 232 OCU at Gaydon in December 1959 there to meet my future crew. The captain was Dick Hayward who was a gentleman of quietest demeanour but ran the crew with discipline and humour. Dick was a confirmed bachelor in those days and had a passion for classical jazz and full strength cigarettes. The nav plotter was Ron Ventham whom I knew on 50 Squadron Canberras at Upwood. He was a highly strung chap whose hobby was flamenco dancing, complete with high heels and sombrero. I believe he once gave me a pair of castanets for my birthday. Our nav radar was Sid Gordon, a tall slim cultured sort of bloke, also ex Canberra. Finally, the AEOp was Master Signaller Doug Smith, a mature knowledgable and thoroughly likeable ex maritime operator.

Episode One – Say what you Mean!
On 18th February 1960 we flew for the first time as a complete crew. It was Dick's first trip as a V bomber captain and only my second sortie in the Valiant. As we came over the hedge at Gaydon for our first roller landing Dick rounded out and said "take off power". In my capacity as officer in charge of throttle I opened up to full power. The situation was thus that Dick was trying to land the aircraft whilst

my action was persuading it not to. The result was the aeroplane accelerated rapidly in the landing attitude; in other words it climbed. During the ensuing discussion, in reality a monologue from the left-hand seat, it was agreed that in future the term 'slow cut' would be used to persuade me to close the throttles, and 'full power' to move them in the opposite direction. This worked well thereafter. Incidentally the throttle levers were a master piece of design. They were about a foot long, delicately curved with a useful relight button on the end. Being on the central console both pilots could caress at will but not both at the same time.

Episode Two – Skipper's Striptease

I am unable to pinpoint the exact date of this saga, but sometime in the summer of 1960, we were over flying France during a routine trip. For some medical reason Dick had had a penicillin jab a couple of days before. Suddenly he became quite agitated and began to scratch at his body. Giving me control, he rapidly unstrapped and disappeared downstairs into the rear crew's domain. To the surprise and unease of Sid, Ron and Doug, Dick stripped off his flying suit and underwear and continued to attack the now very obvious body rash. Not a convenient moment to discover he was allergic to penicillin. I flew the aircraft back to Wittering, though our brave captain did resume his seat in intense discomfort for the approach and landing. This was my first moment of real usefulness, Dick's last penicillin jab and the rear crew's only appraisal of his physique!

Episode Three – Enter the Cavalry 31st August 1960

By now serving on 49 Squadron at Wittering we found ourselves at El Adem, the RAF's paradise in the Libyan desert. At dawn we were about to enter the officers' mess (well everything was a mess at El Adem, including the food) for breakfast, when a dashing figure charged past on a beautiful grey horse. It was going too fast to see closely underneath whether it was a mare or a stallion but we hailed it with cries of "Heigh Ho, Silver" "Where's the Indians" and "is it John Pearce on his way to Widdecombe Fair" and similar clever remarks. A passing officer, resident at El Adem, advised us to desist, as the station commander preferred his morning gallop to be unsullied, so we did. Later that day, raining 100lb practice bombs into the El Adem bombing range, the range controller advised: "Do not drop there are Arabs on the range." Doug was working the radio and asked "whereabouts?" "They are sitting on the target", was the response from the ground. In unison all five of us said, "They are safe then!" I don't know why our turban and sheet clad friends were so anxious to collect sharp lumps of metal but their intrusions were apparently very regular in to the range. Lawrence of Arabia never mentioned this behaviour in his *Seven Pillars of Wisdom*. Perhaps it was in the Eighth Pillar?

Episode Four – The demise of Ron 3rd January 1961

Returning from a lone ranger to Nairobi, we were coasting in over Monte Carlo at 40,000ft and a bit, it was a lovely clear day. A routine call to the French radar services was acknowledged and the helpful controller confirmed our position. Unfortunately he passed us an entirely incorrect latitude and longitude which placed us in the middle of Spain. Ron Ventham plotted this position and lost his cool. He screamed to Dick to turn east rapidly (i.e. 60 degrees of bank) and was sure we would be shot down for infringing Spanish airspace. Nothing Dick said would convince him that we had not, in time warp fashion, jumped from the Côte d'Azur to Madrid. We even told him we could see the totty on the beach at St Tropez. To no avail, poor Ron threw his divider away, narrowly missing Sid's hand, which was reaching for a Mars Bar at the time, saying "I can't cope" and removed his helmet. He was useless for the remainder of the recovery to Wittering and shortly afterwards disappeared from the squadron. Nice chap but strange. John Conning joined the crew to replace Ventham. He was a bundle of fun and we are still in contact.

Episode Five – Back to Front 19th April 1961
During the annual Bomber Command bombing competition we somehow overshot one of the RBS targets before starting the simulated bombing run. Executing an immaculate 180 degree turn, Sid then produced a very accurate attack. Our good result was disallowed as we had bombed from the opposite direction to that in the plan. I always imagined that if you hit a target with a nuclear weapon the direction it came from was somewhat irrelevant.

Episode 6 HHBT – Have had Better Trips
On 16th September 1961 the crew was briefed to fly displays at Valley and Aldergrove, for Battle of Britain Day. Unfortunately Hayward was away having, I think, rolled his Standard Vanguard into a ditch so the left-hand seat and captaincy was occupied by Sqn Ldr Alder who was the flight commander. He was a large, loud, unpleasant individual who despite his mature years suffered badly from acne. Like several V bomber captains he regarded co-pilots as paid hirelings to do his bidding, and rear crew as unnecessary passengers. On this day we duly displayed at Valley, that is flew past very fast, very low and very noisily, then departed for Aldergrove. Over Northern Ireland Conning advised, "Plotter to captain ten miles to Aldergrove." Alder replied, "Shut up nav I have the airfield in sight." So we repeated our enterprising display and whistled down the runway. "Not much of a crowd," said Alder "and what is that Viscount doing on the runway?" John Conning said, "Eight miles to Aldergrove." We departed Belfast civil airport as sharply as possible.

Episode Seven – Gilly makes a Spectacle of Himself
In January 1962 I went back to Gaydon to do the intermediate co-pilot conversion course. This would qualify me to fly the Valiant from the left-hand seat and act as captain under the supervision of a real captain. During the first night trip I was in the tender hands of Gilly Potter, a well-respected and well-liked instructor. He

was demonstrating a night circuit when he suddenly exclaimed, "my glasses have broken". I anxiously offered to take control. "No," he said "I can see the dials but cannot read the numbers. Would you call height, heading and air speed to me?" So Gilly had 80 tons of aeroplane in one hand and his shattered spectacles in the other. He flew an immaculate circuit with me reading the numbers. "Can you see the runway?" he asked. "Yes" I replied, "Good so can I", and executed a faultless landing. The day after returning to Marham from the ICC I was told of my imminent posting on to Victors.

Episode Eight – Put another Shilling in the Meter
On 28th February 1962, we were cruising over the North Sea when Doug announced, "AEO to captain". We knew something serious had happened as Doug would have normally said "Hey skipper". Serious it was. All four generators had gone offline. The Valiant, being an all-electric aeroplane, with no generator output control relied on the battery. Quite like running an electric cooker from a torch battery. Dick asked Doug to carry out load shedding, a technical term meaning switch off damn near everything then attempt to reset the generators. At the time load shedding might have been interpreted as something extremely personal and uncomfortable to sit on. The generators did come back online and we repaired post-haste to Marham.

Episode Nine – 2nd April 1962 – Give us a Lift
On a western ranger we diverted to Ernest Harmon AFB in Newfoundland, a USAF base. During refuelling a US servicing airman wedged the cover to the refuelling point under one of the aircraft tyres. Having uploaded forty-odd tons of fuel, said airman was non-plussed to find that the tyres had squashed a bit. Result – refuelling cover stuck. A very large enlisted man, whose complexion rendered him invisible at night, was then seen to crouch under the underwing fuel tank and attempt to lift the aeroplane by straightening up. This ploy didn't work. A hydraulic jack was obtained and our hero retrieved the offending cover. We thought it was funny even if he didn't.

Episode Ten – 14th June 1962 Farewell
My last sortie on the Valiant was a routine Group exercise, and nothing funny happened. I had flown 480 hours and learnt a lot mainly courtesy of Dick Hayward. He was always generous in giving me pole time and taught me by his professionalism and example how to be a V bomber captain. I tried to emulate him during my many years on the Mk1 Victor. I always considered the Valiant to be an aircraft for gentlemen, the Victor an aircraft for sportsmen with its racy lines and the Vulcan suitable only for hooligans with its fighter pilot stick and excessive noise.

Chapter Eighteen

VALIANT ACCIDENTS
Tony Blackman

It is important when looking at the history and performance of an aircraft to review the accidents that have occurred to see if there is any obvious pattern which indicates a weakness in design. Of the three V bombers it is always said that the Valiant was conventional but that has to be a relative term since in fact it was the UK's first four-jet bomber beating the Vulcan by fifteen months.

Like the other two V bombers that followed the Valiant the aircraft had a 112V bus bar powered by four-engine driven generators. However unlike the Victor and the Vulcan nearly every system in the aircraft was driven from these bus bars and very little reliance was placed on hydraulics. Moreover, there was one important difference between the aircraft and that was in the design of the flying controls; in the event of an electrical failure the Valiant pilots could fly the aircraft since the control hand wheels were connected to the controls while the other aircraft had no manual reversion.

WB210 11th January 1952 Vickers

The first Valiant accident was on this date. It happened to the first prototype WB210 being flown by the Vickers chief test pilot. An inner engine caught fire while it was being relit. Jock Bryce managed brilliantly to get the rear crew out but unfortunately the RAF co-pilot was killed as he ejected hitting the aircraft's tail. Jock's personal account of the whole flight, reproduced in the prologue, ensured that the reason for the accident was clearly understood unlike some of the accidents below. No significant airframe design changes were required as a result of the accident.

WP222 29th July 1955 138 Squadron

On the morning of 29th July, WP222 crashed at Barnack, shortly after take-off in an easterly direction. The accident enquiry concluded that a runaway aileron trim tab actuator caused the accident. The aircraft banked to port at 60° before it struck the ground and exploded. It has been said that on early Valiants the trim switch was very close to the 'press to talk' VHF button and it might have been inadvertently actuated by one of the pilots. However, the Board of Inquiry (BoI) concluded that the runaway was caused by a short circuit in the electrical supply to the trim tab actuator.

The BoI was advised by Vickers that the aileron trimmer switch was double pole but only one of the poles was used to operate the trimmer. The recommendation was to modify the circuit to ensure that both poles had to be selected for the trimmer to operate. Apparently steps were taken to ensure that the trim tab travel was limited on all aircraft. As far as is known the main spars were not inspected for failure due to fatigue.

Top: Mike Beetham leaving Marham for Cape Town. (*Mike Beetham*)

Above: Mike Beetham returning from Cape Town after breaking the record. (*Mike Beetham*)

Opposite top: Mike Beetham's 214 Squadron refuelling top view. (*Mike Beetham*)
Opposite below: Mike Beetham's 214 Squadron refuelling from below. (*Mike Beetham*)

Top: Standard refuelling view. (*Mike Beetham*)
Above: Valiant WZ390 refuelling.

Above: Valiant WZ390 refuelling a Javelin on the way to Gan. (*Alan McDonald*)
Left: Tanking Valiant.
Opposite top: Aircrew caravan on a dispersal airfield. (*Associated Newspapers Solo Syndication*)
Opposite bottom: 148 Squadron Valiant line up. (*Anthony Wright*)

Top: 49 Squadron in 1961. (*Keith Walker*)
Above: Valiant crew scrambling to launch their aircraft to deliver a nuclear attack behind the Iron Curtain. (*Courtesy of Martyn Chorlton, Old Forge Publishing*)

Opposite top: Morning Depature. (painting by *David Wright*)
Opposite bottom: Duke of Edinburgh talking to Dave Roberts's crew. (*Jo Lewis*)

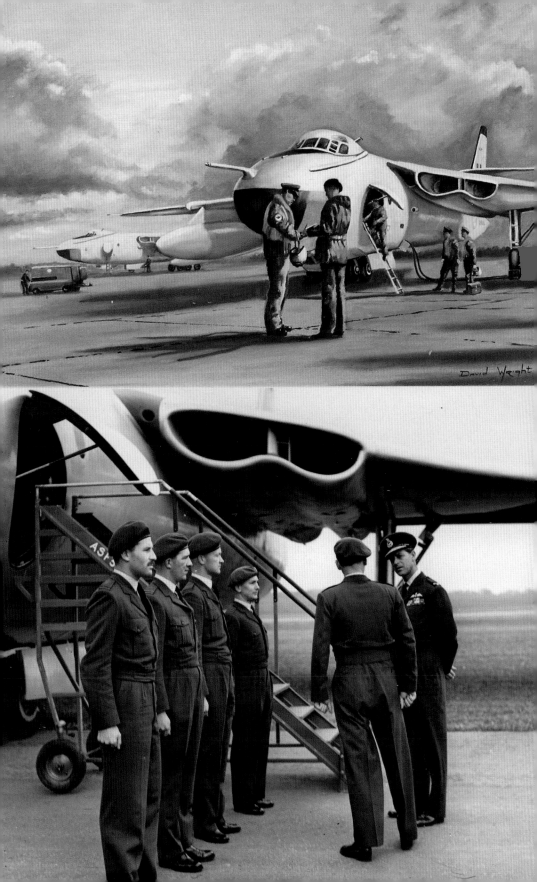

Top: Snowy take-off at Marham.
(painting by *David Wright*)

Above: Flight refuelling at dawn.
(painting by *David Wright*)

WP202 11th May 1956 Royal Aircraft Establishment (RAE)

Suffered a major electrical failure at low level and high speed south of Farnborough. The pilot attempted a crash landing near Southwick, Sussex but the aircraft exploded when it hit the ground. Of the four crew, only the co-pilot survived when he ejected moments before the aircraft struck the ground. Five civilians were also injured.

With the failure of the 112V supply the controls would have reverted to manual control and in theory it should have been possible for the two pilots together to control the aircraft; however despite all their efforts this proved to be impossible. This point was not discussed by the BoI though the tail plane setting was examined and found to correspond to 300kts which was their estimated speed.

This accident must have been very concerning to Vickers and all Valiant operators since there should have been no electrical failure that would cause the aircraft to crash. Mr Harry Zeffert, Vickers' chief electrical engineer, gave evidence to the BoI and said Vickers were "active to the implications of the accident". If the two pilots could not control the aircraft one has to suspect a trimmer at full travel but apparently this was not the case.

WZ398 13th September 1957 543 Squadron

While work on the fuel tanks was in progress in a hangar at Wyton, Cambridgeshire, there were live 112V electric power contacts exposed due to airbrake servicing. An explosion occurred in the bomb bay and the aircraft was destroyed by fire.

XD869 11th September 1959

Crashed in a tail down attitude four miles east of Marham, Norfolk shortly after a night take-off for a direct flight to Nairobi; the power was heard to decrease and then increase just before impact. All six crew were killed. The aircraft had not been test flown following twenty-two days of servicing and the accident was possibly caused by the tail plane incidence switch working in reverse. After take-off the pilot would have operated the switch to select nose-down trim as the flaps were retracted and the BoI suspected runaway trim in the wrong direction. Inspection of the wreckage showed that the tailplane was set at 1.5° nose down whereas it would have been expected to be at 0 to ¾° nose up. The flaps were found to be set at 3°. There was no evidence given by a Vickers electrical engineer. As far as is known the main spars were not inspected for failure due to fatigue.

XD864 9th August 1960 7 Squadron

This accident occurred at 10.38 BST after taking off from Wittering. The known facts were that the aircraft made an unusually steep take-off and the nose wheel failed to retract; the aircraft then turned left after take-off by at least 120° and crashed on Spanhoe disused airfield, killing all five aircrew occupants of the aircraft.

Like all non-test Valiants there were no flight data recorders and the BoI was faced with determining the cause of the crash. They were helped by a civilian accident investigator from the Air Accident Investigation Board and the board interviewed many eye witnesses to try to determine what actually happened. Apparently the aircraft was clean apart from the nose landing gear and the aircraft appeared to be going very slowly as it turned. Abrupt wing drops were seen and there did not appear to be a lot of engine power. The board then used a flight simulator to fly the flight path described by the witnesses and came to the conclusion that the aircraft had stalled uncontrollably, hitting the ground.

It was believed that the co-pilot, not a regular crew member, did the take-off and achieved too steep an angle. The board felt that the flaps were retracted too early after take-off and this would have definitely contributed to the likelihood of a stall. The findings were:

Primary Cause of the Accident.

The Board find that the primary cause of the accident was pilot error in entering a turn with insufficient speed and/or engine power to maintain an adequate safety margin over the stalling speed for the relevant flight configuration, i.e. clean aircraft. This resulted in loss of control and the crash of the aircraft.

Secondary Causes of the Accident.

The Board find that the secondary causes of the accident were :-

(a) The failure of the nosewheel to retract.

(b) Pilot error in retracting the flaps before complete retraction of the undercarriage. This is contrary to standard procedure as taught and as detailed in A.P. 4377 A. - Pilots Notes for the Valiant Aircraft, Part IV, Paragraph 10.

Findings of the Board of Inquiry.

My only comment is that there were broken spars in the wreckage and apparently no tests were carried out to determine whether the failure was due to hitting the ground or whether the spars showed any signs of fatigue bearing in mind that it had been known for four years that the main spar material had a very low fatigue life[13].

It is always very easy to blame the crew when investigating an accident if they are not available to comment. Such blame relieves the manufacturer of any primary responsibility and also other agencies like maintenance and operational staff. Currently aircraft investigation bodies are much more careful before blaming the crew and they have crash recorders to help them but certainly in the past there have been many accidents on different aircraft where a little less certainty on 'crew error' might have been desirable.

---

[13]   See next chapter.

WP205 17th November 1960 A&AEE

Overshot the runway on landing at Boscombe Down, Wiltshire and struck the control caravan. Not repaired and struck off charge on 11th July 1961. It has proved impossible to ascertain the circumstances prevailing which led to this accident.

WP200 14th March 1961 Royal Radar Establishment (RRE)

A take-off from RRE, Pershore, Worcestershire, was abandoned after the powered controls were inadvertently switched off while the crew were attempting to clear a restriction in the port rudder movement. It overran at high speed, crashed through a steel mesh fence and came to rest just short of a railway line. Two crew were severely injured, the navigator being trapped in the cockpit for four hours. The other crew member received minor injuries. Again it has proved impossible to find the BoI for this accident but it certainly seems that in this case there was an element of crew error.

WZ399 3rd November 1961 543 Squadron

This is an accident where the crew survived and so the details are known. During a high weight take-off from Offutt AFB, Nebraska, USA both airspeed indicators failed at or very close to rotation speed and the captain decided to abort the take-off. The main wheel tyres burst and the aircraft overshot the runway. Due to the ground falling away it became airborne and cleared a six-lane highway before crashing onto a railway line. The crew survived when the cockpit broke away and was catapulted across the railway line away from the burning fuselage. The primary cause of the airspeed indication failure was due to the wrong covers being fitted over the pitot tubes during the night; this allowed water into both air speed indicator (ASI) systems and then during the take-off the water froze causing the ASI failure.

We are fortunate to have a personal account of the accident from **Bill Yates**, nav plotter who unfortunately was injured in the accident.

It was Friday 3rd November 1961 and we were planning our return back to our base RAF Wyton. We had been at US AFB Offutt for five days carrying out radar training along the US and Canadian border. The weather factor had not been good due to several deep depressions over the central and northern parts of the USA. We had strong winds and very low temperatures. The previous day, our sortie was cancelled due to bad weather, so the aircraft was put to bed on the disused runway. During the night another weather pattern went through the area with below zero temperatures.

It was our habit always to get airborne on time to the exact minute, so after met briefing I completed my route plan back to Newfoundland on the homeward bound leg, and the flight plan was filed with ATC with our take-off time as 0800

Aircraft WZ399 on the railway; all the crew survived. (*Bill Yates*)

local time. From there we took the flight ops coach to the aircraft, it had been prepared by the ground crew so the only thing the captain or the co-pilot had to do was make a visual check of the outside of the aircraft. We all climbed aboard, closed the cabin door, went through our checks and then taxied on to the runway for take-off.

During the night the temperature had dropped, and despite this we didn't know what was to follow. The pilots did the final pre-take-off checks, on to max power, brakes off, the runway was long about 10,000ft, checks were made on the ASIs and the take-off speed was in the region of 170kts. By the time max power was reached, one air speed indicator was fluctuating and it may even have stopped. We had reached our time to lift off, but then the other pilot's ASI went the same way as the other. You must realise the captain had to make an instant decision – abandon take-off or continue with take-off. He decided to shut down all four engines and apply maximum braking; travelling at over 200 miles per hour and the aircraft with max fuel on board.

All the main tyres then burst, we were still at a possible take-off speed when we reached the end of the runway, and what did the men up front see beyond the runway? To their horror they saw that the ground fell away, a possible thirty degree downward slope, and at the bottom a six-lane highway.

The aircraft did get airborne and then hit the ground. All the equipment in front of me was on fire, we cleared the highway still airborne and the next impact was on a minor railway line which finally caused the aircraft to break up and burn out completely.

At the BoI the pilots were found responsible. One small thing had triggered the accident, unknown to the crew, the pitot head covers used overnight (low freezing temperatures remember) were of the wrong type, which allowed water to get into the ASI system.

It was later discovered that ice was seen in the pitot head, and the ground crew whilst putting power to the aircraft during their checks melted the ice, then it drained down into the air speed mechanism. Whilst we were taking off with the minus temperature, it re-froze which caused the ASI to fluctuate and then failed with catastrophic results.

The pilots were blamed in that the decision was taken to abandon the take-off when above the speed when it was possible to stop on the runway, V1. The Valiant simulator crew drills apparently did not include a double ASI failure after V1 but it is considered that the pilots would have been able to take off since there would have been plenty of time to find a formatting aircraft to advise the crew on approach airspeeds.

Unfortunately for Bill even after hospitalisation, treatment and rehabilitation over three years he lost his flying status, and twelve years later his flying was terminated with the loss of his flying pay. He eventually retired aged fifty-one and then took up a job he found satisfying in ATC, and, with the combination of flying and air traffic, joined civil aviation instructing airlines in navigation and aircraft performance.

WZ363 6th May 1964 148 Squadron aircraft on loan to 207 Squadron

At 23.50 WZ363 crashed into farmland near Market Rasen in Lincolnshire. Once again the BoI was put in a very difficult position because there was no crash recorder. The aircraft had been working air traffic quite normally using first local at RAF Marham and then approach control with primary radar. The controller saw the blip disappear off his screen which at first did not alarm him but then he enlisted help; shortly afterwards the crash was reported.

Inspection of the wreckage showed that the aircraft had dived into the ground at very high speed and after detailed inspection of the wreckage it was discovered that the tailplane actuator was in a fully nose down position. The BoI came to the conclusion that the trimmer must have malfunctioned. A runaway trimmer was suspected due to an electrical malfunction and it was discovered that there had been one electrical malfunction of the trimmer previously in another Valiant when the trimmer moved in one direction only regardless of the direction selected by the pilots.

The BoI felt that the trim speed would have been set at fast/coarse and if the trim started moving in the wrong sense it would only take a few seconds for the trim to be fully down. The pilot's instinct would have been to try at least once more to trim the correct way which would have made matters worse. Furthermore the pilot might have asked the other pilot to help and this pilot would have inevitably tried trimming exacerbating the situation so that with both pilots pulling at the control column they would have been unable to prevent the aircraft diving into the ground. The fact that it all happened at night would not have made things any easier.

Clearly the whole event happened very quickly with no Mayday which would be very understandable since it is likely that both pilots were fully committed trying to prevent the aircraft from diving.

It has not been possible to find the cause for this accident and the accident could have been due to something entirely different. It is not known whether the spars were examined for fatigue but it must be borne in mind that the accident occurred only three months before the spar failure of XP217 on 6th August 1964.

WZ396 23rd May 1964 543 Squadron

The starboard main undercarriage door came open during a flypast at Bentwaters, Suffolk. It broke away preventing the wheel from being lowered. A wheels-up landing was made at Manston, Kent on a foam carpet but the aircraft was written off.

WP217 6th August 1964 232 OCU

Suffered a wing spar failure over North Wales. It made a safe return to Gaydon, Warwickshire. After this incident all other Valiants were inspected and many were found to have defective wing spars. This resulted in the early withdrawal of the whole fleet.

Looking at the Valiant accidents it was of course quite impossible for the BoI to come to any definite conclusion on some of the tragedies because there were no recordings of the instrumentation that survived the impacts. However, there were at least three cases where the trimming system appeared to be at fault. This I found slightly surprising since when I was at Boscombe in 1955 and 1956 there had been at least one trim runaway and we spent a lot of time on all the aircraft we were accepting ensuring that they were modified so that the electrical trimmers were all double poled in operation; Harry Zeffert, who I met, was a very well respected Vickers electrical engineer and it would be surprising if the Valiant aircraft were not all modified at that time.

The other point was that the main spars of the Valiant proved to be liable not only to fatigue cracks but also to loss of strength over the passage of time because of the characteristics of alloy used, DTD683. There had been an early spar failure on the RATOG test aircraft described in Chapter Nine. As far as could be determined the spars of the aircraft that crashed due to trimmer malfunction were not inspected for fatigue nor was the aircraft that made a steep climb after take-off and then was judged to have stalled. It is considered that there has to be some element of doubt on the primary cause of some of these accidents. Fatigue is discussed in more detail in the next chapter.

Chapter Nineteen

THE SUDDEN FINISH –
A SURPRISE TO SOME?

When I (Tony Blackman) started writing this chapter I found it difficult to square various events, particularly after I understood the history of the metal used in the Valiant's spar.

In August 1964 when WP217's rear spar broke the RAF's reaction was immediate; all the other Valiants were inspected, cracks found and as a result the aircraft were grounded. Apparently the event was a complete surprise not only to the operators but to the whole of the RAF. However three months earlier, a Valiant when carrying out a photographic survey in Rhodesia was found to have a rear spar that was cracked. Furthermore, in 1957 the rear spar broke in WP215 when carrying out RATOG trials at Boscombe Down and the break was already put down to fatigue by a Vickers team.

Conclusions

The conclusions revealed during the initial inspection (WMT 12/1-430) were substantially confirmed by this examination. Primary fatigue initiation had occurred on the rear face of the rear web adjacent to station 106.5 at the innermost liner attachment bolt hole. The rapid failure from the initiation below the bolt being stopped by an adjacent rivet hole may have prevented total failure occurring at an earlier date.

Compiled by: H. Tyrer

Approved by:

The Mechanical Test Laboratory,
Vickers-Armstrongs (Aircraft) Ltd.,

Report on WP215's broken rear spar.

One has to assume that not only the engineers at Vickers, but also those in the RAE and the RAF, must have been aware of the dangers to the whole Valiant fleet at that time; and the year before WP215's cracked spar in 1956 there had been two widely circulated articles pointing out the poor characteristics of the alloy being used for the Valiant spars, DTD683. The first article was presented by Lockheed's chief structural engineer at the RAeS in 1956[14] and the second by T Broom was published in 1956 giving details of the structural changes to DTD683 caused by

---

14 McBrearty J F, 'Fatigue and Fail Safe Airframe Design', *Society of Automotive Engineers*, 1955; *Flight* 6 April 1956, p394 (RAeS lecture).

167

plastic strain and fatigue.[15] The shortcomings of the alloy were clearly no surprise to Vickers as in 1951 the assistant chief designer Mr H H Gardner described the problems they encountered and the engineering techniques they had to employ to manufacture the aircraft in a talk to the RAeS on 29th September 1951 and reported in *Flight* magazine[16]:

1) Distortion after machining, partially solved by machining before age hardening and after a 'controlled stretch' of the material, Mr Gardner says this was an unsatisfactory aspect of the material;

2) Variation in strength across the section, which gave low core properties;

3) "The lack of stability was an unpleasant feature";

4) Variation in strength between longitudinal and transverse grain directions resulting in "all-too frequent failures" (referred to in the literature as shear failure);

5) High (internal) stress concentrations.

This quote from Mr Gardner's 1951 lecture also foretold the Comet disaster in 1953:

"... since most pressure cabins had to carry main structural loads as well as pressure loads, the possibility of fatigue failure was an anxiety ..."

Despite the warnings in the technical articles and the actual breakage of WP215's spar, any regular inspections for fatigue damage seemed to have been very ineffective since, when WP217's spar actually did break and proper inspections were carried out, most of the aircraft were found to have severe fatigue damage.

Inspection of RAE records in 1957 and later show that Vickers set up a fatigue test specimen at their facility at South Marston and requested authority to proceed but apparently the cost spiralled and it is not clear whether the tests were ever carried out. One suspects not because in 1963 there was a paper in the OR Branch Bomber Command examining the difference in fatigue life of the Valiant if it were to operate low level rather than high, concluding that the fatigue life would be reduced by three quarters operating at 1,000ft and halved again if operating at 500ft. In addition the same paper concluded:

---

[15]   Broom T, Mazza J A, Whitaker V N, '1790 Structural Changes caused by Plastic Strain and by Fatigue in Aluminium-Zinc-Magnesium-Copper alloys corresponding to DTD683', *Journal of the Institute of Metals*, Vol 86, Issue 8-13, 1956. Further information on this article can be found: http://zkt.black fish.org.uk/XD864/

[16]   Gardner H H, 'Structural Problems' – RAeS Lecture 29 November 1951, *Flight*, 14 December 1951, pp756-758.

'If the Force is expected to last until 1968 the fatigue life could become critical should the training programme be extended without a careful examination of the rate of use of fatigue life at all times.' And 'Using this very limited sample of data, prediction of the life of the Force for different training programmes has shown that the fatigue life could become a limiting factor unless carefully checked at all times and with each change of training requirement.'

It is quite clear that Vickers used an unsatisfactory metal alloy for the Valiant which gave the aircraft a low fatigue life; the alloy was almost certainly specified by the Ministry of Supply and called in the UK DTD683, aluminium with some zinc, magnesium and copper. Perhaps at the time the shortcomings of this particular alloy were not appreciated when using it for a high altitude jet that was not in immediate danger from warfare and would therefore be expected to have a long time in RAF service. The choice of this alloy was particularly unfortunate as the aircraft was designed on a 'safe-life' principle as distinct from 'fail-safe' so that airframe failures tended to be abrupt rather than a gradual decline in strength. Despite all the knowledge of the shortcomings of DTD683, when the spar failed in 1964 and the decision was taken to ground the Valiant the following paper was sent to Harold Wilson the then prime minister saying:

The Fatigue Problem

7.   In August 1964 a fracture appeared in the rear spar of a Valiant.   This was found to be due to metal fatigue.   As a result the Valiants were inspected and indications of fatigue were found in nearly all of them.   The aircraft with the most serious cracks were grounded and the others were allowed to fly under certain stringent restrictions.   At this time it was thought that, of the total R.A.F. fleet of 61 aircraft, perhaps 40 could be repaired.   Subsequently, however, cracks were found in the front spar of one aircraft and also in its rear spar outside the area that had previously been inspected. The entire Valiant Fleet was accordingly grounded on 9th December and we started a thoroughgoing assessment of the wing structure of the aircraft

Brief to the prime minister on Valiant grounding.

This brief from the RAF to the government and the prime minister infers that the rear spar fracture was a complete surprise which in the circumstances seems absolutely amazing. Indeed from Mr Gardner's remarks in 1951 we know that there were people at Vickers who realised that the life of the Valiant would be very short indeed and it is difficult to believe that MOD senior engineers were not aware of the problem; the situation was particularly important because the nature of the alloy meant that the failure of the spar could be very sudden. It is believed that strain gauge counters or fatigue meters were introduced at a late stage but it is not clear how the information was used. When the decision was taken for the UK deterrent to go low level and include the Valiant it is hard to believe that there were not people in the corridors of power who were advising against such a role change and conversely there must have been people who, despite knowing the fatigue situation, decided to ignore the advice.

Fifty-eight years have passed and it is most unlikely that it will ever be established who knew what and when. If there was a regular inspection of the spars it must have been very ineffective since only after WP217's spar broke were all the other Valiants immediately checked and the damage established.

Anthony Wright's first sentence below uses the phrase 'fatigue was discovered' which supports the view that the routine operators were apparently surprised at the event. However, the signs of fatigue must have been clearly visible before XP217's rear spar broke since all the aircraft were immediately classified according to the extent of fatigue damage as indicated in the brief to the prime minister.

In his book *V-Bombers*, Barry Jones notes[17]:

'A question has to be asked. For two years before the demise of the Valiant, Handley Page at Radlett had 100 Hastings go through their shops. They were completely dismantled and rebuilt, having DTD683 components removed and replaced by new alloy sections. What was so special about the Hastings and why was the Valiant not treated similarly? Perhaps we will know one day – but I doubt it.'

Clearly the Valiant was very important at that time in supporting the nuclear deterrent policy and the Government wanted the Valiant to keep flying as long as possible but with the Vulcan and Victor not far behind it almost certainly did not seem worth incurring the cost of resparring. However, as can be seen in the previous chapter, it is possible that accidents due to fatigue failure of the spars might have occurred as a result.

---

[17]   *V-Bombers: Valiant, Vulcan and Victor*, Barry Jones, The Crowood Press, 2007, p117.

Anthony Wright relates what it was like to be on a Valiant squadron as the aircraft were grounded.

It was on 6th August 1964 that metal fatigue was discovered in the Valiant fleet by my old captain, Taff Foreman, flying an instructional sortie at Gaydon when after experiencing a large bang and airframe shudder he was forced to land. Among other things, it was discovered that the rear spar of the starboard mainplane was cracked. The ramification of this was that it eventually caused us to undertake reduced flying. Although the Valiant was temporarily grounded, shortly after, I was flying them again starting on 18th August although we were restricted as to the distance that we could fly from base. All we knew was that bombing runs could only be carried out on nearby RBSUs and then only from certain directions. No longer could we expect the luxury of 120 miles run in to a high level target. Everything was now geared to preserve any fatigue life in the aircraft. Meanwhile, the Vickers teams were working on the problem and by October had the Valiants placed into three different categories namely: Category A — fit to fly, Category B — able to fly in an emergency and Category C — grounded.

I continued to fly throughout September, October and November. Although that last month I flew just five sorties each of which was a nearby RBSU such as Kenley, London, day and night, and just continuation training flying in the circuit to keep the pilots in currency. The last time that I flew in a Valiant was on 25th November. There were moves afoot to repair those Valiants affected. However, it was soon discovered that the front spar was also affected. Therefore, on 9th December all Valiants were grounded, and only to be flown in a national emergency. Our QRA commitment both national, and to SACEUR of course, continued as normal. By now the writing was on the wall for the Valiant and it would only be a matter of time when the axe would fall. Of course only a cynic would link the fact that Churchill died on 24th January 1965, and lay in state for three days, the funeral was on 30th January and in between these dates came the breaking news of the end of the Valiant. With the national newspapers devoted to Churchill's death, what better time to bury bad news?

For us the final crunch came on the evening of 26th January when I was on QRA. My nav plotter, Sqn Ldr Ken Lewis, and I weren't watching TV but playing snooker when Jo, his wife, rang up to tell us that we were redundant. She said that it had been just announced on the evening news that the Valiants were to be withdrawn from service which would take effect the next day. Shortly after the news the station commander, Group Captain Kennedy, who had only been informed 10 minutes before the announcement, arrived along with the three squadron commanders (OC's 49, 148 and 207 Squadrons) of the crews on QRA, with crates of beer. He told us that if there was a call out we were not to react. The QRA aircraft would be downloaded the next day. The defence correspondent of *The Times* reported the news on Wednesday 27th January. An extract of three of the paragraphs from the article says it all:

'The complete Valiant force of Bomber Command is to be taken out of service forthwith, the Ministry of Defence announced last night. All Valiants have been grounded since December 9. Investigations into cracks in the wings of some bombers disclose that the whole fleet was suffering from metal fatigue. It has affected the main spars in the Valiant's wings, front and rear.

'Mr Healey, Ministry of Defence, intended to make the statement to the House of Commons yesterday, but this plan was superseded by the adjournment of the House in deference to the death of Sir Winston Churchill. Instead, Mr Healey will be prepared to answer questions on the statement next Monday, when Parliament resumes.

'The number of service men engaged on Valiant operations is not disclosed, but the RAF have no qualms about their easy absorption into other branches of the 20,000 strong Bomber Command. The airfield at Marham, Norfolk, which has been an exclusive Valiant base for some years, may be closed.'

Although we knew that the withdrawal of the Valiant had to happen sometime it still came as a bit of a shock. And in true aircrew tradition having whinged about the Valiant and our lot in the V Force, from time to time, everyone now sat around beer in hand saying what a great aircraft it was and what good times we'd had! The next subject that would be constantly on our minds for the next few weeks or months, was what was going to happen to us. Luckily for the RAF we had a few little wars and skirmishes going on at that time. Indonesian Confrontation with postings to Labuan and Kuching in Borneo, unaccompanied year-long postings to Sharjah in the United Arab Emirates (UAE) and postings on helicopters in Aden were just some examples for which there was a list in the crew room. Many didn't volunteer for Borneo or for unaccompanied year-long postings to the UAE. The other downside was that if husbands were posted on unaccompanied postings their wives were housed in surplus quarters situated on disused RAF stations away from friends and families. I volunteered for helicopters by appending my name to the list. My nav plotter, so called father figure, said, on seeing my name, that it wasn't a good career move and crossed it out. It was the days when flying officers didn't get away with arguing the toss with squadron leader flight commanders and I just had to await my fate. There was no such thing as negotiation with postings unless they wanted you to volunteer for something. Mind you if they got no volunteers they just went ahead and volunteered you. My AEO ended up in Sharjah.

With so many aircrew and ground crew on 49 Squadron, 148 Squadron, 207 Squadron and 214 Squadron at RAF Marham, on 90 Squadron at RAF Honington and 543 Squadron at RAF Wyton, with virtually nothing to do, it was quite a feat to keep everyone occupied.

After the aircrew and ground crew at Marham had exhausted playing each other at various sports, it was decided to keep the aircrew in line by compulsory Met briefings every morning at 8am. Everyone assembled in the Operations Wing briefing room to hear the weather forecast that we had just read about in the morning papers. This was followed by any notices and then to keep us on our toes all junior officers were ordered to give ten to fifteen-minute talks on a pet subject to the rest of the 220 plus aircrew. Stage fright was not an option. I gave my talk on investing in stocks and shares from a book that I'd bought, and mugged up on, at a W H Smiths bookshop . It must have had some impact as a squadron leader came up to me afterwards to ask my expert advice.

After the talks a number of thirty-two-seater coaches from MT appeared outside the operations block to take us off on daily trips to various establishments throughout the UK. A visit to RAF Wittering to observe RAF engineers and armourers service Blue Steel weapons was one, a visit to No 1 Aeronautical Information Unit (AIDU) at RAF Northolt was another, while a tour of The Admiralty Compass Laboratory at Slough seemed to include every compass known to man and finally a visit to JARIC were just four, of the many that we endured, that I recall. We were all coach weary by the end of each week.

However, realising that this couldn't go on indefinitely our leaders, as a bit of light relief, brought in some extra Chipmunks on the station establishment. These were for the co-pilots to fly. However, we interpreted that to mean 'let your hair down chaps' and us young nav radars who were equally keen to cause mischief had a great time accompanying them. Flying Officers, John, Galyer and Turner were the three pilots that I teamed up with to have some fun local flying. This meant flying extremely low, up and down the straight fenland dykes, hopping over any bridges, annoying farm workers in the fields and then climbing to finish off with some aerobatics. To supplement the Chipmunks the hierarchy also brought in one Anson for us to fly. This was too tame for most and so as I was Anson qualified, from my two previous spells on Communications Flights, I navigated my Valiant captain, Norman Bevis, just the two of us all over the UK. He subsequently got a posting to Vulcans and desperately wanted me to go along with him as part of his new crew. As I had already experienced one of the V bombers I knew that the job on Vulcans, albeit at a different station, would be exactly the same. Therefore, I turned down his kind offer to await my fate.

Clearly, someone at High Wycombe with common sense for once, was aware that a number of aircrew would end up on Vulcans or Victors. Therefore, in the interim it was decided that it would be useful for some to keep their hand in flying V bombers. To achieve this the plan was for them to go and fly the Victor at 232 OCU. I didn't volunteer, I was just one detailed to go as was the way in how squadron life operated. I duly travelled to RAF Gaydon and spent one morning being briefed on the Victor electrics followed by escape drills in the crew escape trainer in the afternoon. The next day I flew as the nav radar with four other 232 OCU Victor aircrew, captained by Flt Lt Hastings, attacking high level RBSU targets on a four-hour sortie. As the radar kit was exactly the same as a Valiant I found it was no problem. What is more, as an added bonus, I had much better visibility than in the Valiant as I could swivel my seat round and look over the heads of both pilots.

One perhaps understandable shortfall as a result of the sudden grounding of the Valiant was that the squadrons were not closed down formally with parades but just allowed to wither away. There weren't any formal disbandment ceremonies or social activities as such. We were all effectively 'air brushed' out of existence.

> When the decision was taken to ground the Valiant the government tried to recover some money by selling most of the aircraft for scrap. A few aircraft went to airfields for firefighting training and only one airframe was kept complete, XD818, which resides at the RAF Museum at Cosford. Understandably but disappointingly it is not possible for visitors to see inside the flight deck which is complete. We are very lucky to have this airframe since, as Peter Sharp explains, it clearly wasn't a decision taken by the heavy breathers.
>
> **Peter Sharp** was specialist weapon ground crew and he started working on Valiants on 138 Squadron at RAF Wittering in 1960. He went on to 49 Squadron which subsequently moved on to Marham.

I was on Valiants until their unfortunate and premature demise, after which I was posted to Khormaksar, Aden. After my Aden tour, I returned to Marham where the Victor K1 was scheduled into service. We struggled on with a single K1 for some time until sufficient K1s had been modified to make up a consolidated fleet, and then the Tanker Force was born together with 55 and 57 Squadrons. Later I was posted to Akrotiri with the Vulcan build up in the Med. Many years later I was again posted back to Marham, this time to Canberras and Victors and also became involved with the introduction of the Tornado into service.

Thanks to Peter Sharp we can see Valiant XD818 at Cosford.
(*RAF Cosford Museum*)

It was at this point that I once again came across XD818, which at that time was a designated training airframe 7894 M, which it never took up, always displaying her original number. Being an 'old Valiant man' and ex-49 Squadron as well, I was very keen to once again view the cockpit, so I contacted the then custodian who opened the door for me to view the chaos within. She was in an absolutely disgusting state. Apparently a film crew had used her for some filming and had opened up the emergency exit on the starboard side, and the ditching exit in the roof and had not replaced them correctly. Consequently the bomb aimer's blister was flooded to a depth of some twelve inches or so with water. I made immediate moves to gain custody of her which led to an awful lot of problems, but in a nutshell she was on admin wing's charge and I was tech wing which meant that she would have to change wings. The OC engineering wing didn't want that responsibility and his exact words were "so then chief, what happens when you get posted and I am lumbered with her then?" With such a lack of sentimentality and a devotion to making life easier for himself, that I had never directly witnessed before, I was a bit taken aback by his attitude. However it appeared that we had at one time attended the same Grammar school which he started talking about. Somehow we got to the masters and subjects that they taught and I happened to mention the English master, whereupon he stated talking about the books that he had studied for the O-level – which were the same as mine – so we ended up discussing *The Memoirs of a Fox Hunting Man*. To sum up, his attitude changed and he approved the move and all of its implications.

A three-year refurbishment programme then commenced. The rear crew position equipment consisted mainly of full scale photographic images of the items that had originally been installed. Looking back to the period when the aircraft was written off charge, such kit would have been state of the art and still on the secret list, hence its removal. The flooded lower bomb aimer's position though proved to be something of a headache as it seemed impossible effectively to drain all of the accumulated water out. After re-fitting the disturbed hatches, I reluctantly drilled a ¼ inch hole from the outside and drained it that way. The seats were all removed to the ejection seat bay at Marham and serviced, where it was found that although the operating springs had been cocked for some twenty-odd years or so, they were all found to be still within limits when tested. The seats would have functioned as designed, which is very much to Martin Baker's credit. Obviously most of the software fittings (seat belts, dinghy packs, and other bits and pieces) had suffered badly and had to be replaced, but that was no problem, as we had type 2 seats on the Canberras and type 3 seats on the Victors, so parts were readily available. Even the aircrew P tubes were replaced with Victor ones, one of which I still have today. When the Vulcan was being scrapped just prior to the Falklands campaign, I took a three-ton truck up to Waddington and collected all of the electronic kit that was missing from the radio crate of XD818. This was all installed by me to complete the re-fit. I had just replaced the desk tops, the originals having delaminated because of the leaking ditching hatch. Suddenly overnight she had become the star of the show and she was now going to Cosford for permanent display.

Not long after I took charge of her, we had a station open day and I was asked numerous questions about the history of the aircraft. This I knew very little about apart from the fact that she had dropped one of the bombs at Christmas Island, so I decided to dig up what I could starting with the Air Historical Branch. Air Commodore Probert ret'd, then head of this body, invited me to peruse all of the data that they had on record and I duly elected to spend a few days doing so. However upon my arrival on the first day, all of my note pads and stationery were confiscated and I was told that the only information that I could print / publish was what I could remember, which in my case wasn't a lot. So I came back very disappointed and really fuming that I had not been able to do the job properly.

Besides XD818 at Cosford, there are two cockpit sections surviving. XD816 at Brooklands Museum and XD875 at the Highland Aviation Museum at Inverness Airport. There is a third surviving cockpit, XD826, which is part of a private collection in Essex and there are flight decks displayed at the Norfolk and Suffolk Aviation Museum at Flixton, XD857, and at the Brooklands Museum, XD816.

FINAL WORDS

This book is a fascinating look at the birth of the UK's first four-jet long-range bomber. Vickers did a wonderful job in both the aerodynamic and system design of the aircraft but understandably the design was spoilt by the use of the alloy DTD683 for the spars. At the time the decision was made there was insufficient knowledge of the stability and fatigue characteristics of the metal and it must have been very difficult for Vickers to know how to respond as the metallurgists explained the shortcomings of the DTD683.

However the actual operators of the aircraft knew nothing of the design problem and steadily developed the aircraft as a strategic bomber, a reconnaissance aircraft and a flight-refuelling tanker. The major milestones in the life of the aircraft were clearly the Suez campaign, the atomic weapon trials, the reconnaissance role and the development of flight refuelling turning Alan Cobham's dream into reality.

Ironically perhaps the aircraft's greatest contribution was not in the actual flying but in just being there on the ground ready to take-off loaded with a nuclear weapon on a one-way mission to behind the Iron Curtain. The aircraft and the crews recorded no flying hours and it was a great tribute to the operators of the squadrons providing the deterrent that the morale was kept up during the many hours of tedium.

Unlike the Vulcan there is no flying airframe today to keep the country's awareness alive of the Vickers Valiant but hopefully this book will play its part in recording and maintaining the memory of a splendid aircraft.

Tony Blackman
July 2014

NUCLEAR RADIATION – AN UNFINISHED TALE?
Tony Blackman

Notice outside Maralinga. (*Australian National Archives*)

The Legacy of UK nuclear testing

From an operational viewpoint, in both Australia and the Pacific, the nuclear weapon trials were carried out satisfactorily. However there is still a very strong body of opinion that insufficient attention was paid at the time to protecting those present at the sites when explosions took place and in the aftermath when dealing with dangerous surface radioactivity. This book is about the Valiant and the people who operated the aircraft but I felt it was important to mention the alleged effect on the people who supported and enabled the nuclear tests to take place.

The UK government understandably at the time wanted to produce an effective atomic weapon as quickly as possible and the people charged with enabling this to happen seem to have ignored some of the essential protection measures. What is disappointing, to say the least, is that no special effort seemed to have been made by the UK or Australian governments to help the many people present at the tests who developed cancer and other associated radiation problems.

There can be little doubt that the radiation protection of the workers both in Australia and Christmas Island fell far short of current requirements and in composing this appendix I was able to speak to an RAF NCO who was present at Maralinga and Christmas Island during the tests. Both the UK and Australian governments have spent an enormous amount of time and money trying to prove that the cancers and illnesses suffered by the claimants in the court cases were not statistically proved to be due to an excessive amount of radiation and, latterly, by stating that it is now too late to bring court cases due to the length of time that has passed. It is considered likely that both governments are being particularly careful because of the possibility of people who have been exposed to radiation having children with abnormalities which might be blamed on a parent who had been exposed to radiation.

This appendix just mentions a few of the issues and examples but it would need a book to chronicle the material that has been written on this subject.

Maralinga

Dealing first with the use of Maralinga, it is remarkable by present-day standards that authorisation was given by the Australian government to the UK to use the site at all. First of all it was known that prior to selection, the Maralinga site was inhabited by the Pitjantjatjara and Yankunytjatjara Aboriginal peoples, for whom it had a 'great spiritual significance'.[18] However, despite this, the site was approved and apparently many of the Aboriginal people were relocated to a new settlement at Yulata, followed by attempts to curtail access back to the Maralinga site which were often unsuccessful.

Besides the impact on the Aboriginals living in the area, work was carried out on the site by the Australian army with the soldiers being posted to the Task Force Weapons Trial unit. It is quite clear that a lot of the workers, now classified as Australian veterans, were exposed to the radiation from British nuclear bomb testing and got radiation sickness.

A typical case is **Meagan Kopatz's** father's details posted on the internet 8th March 2013:

Between 1952 and 1963, the United Kingdom conducted twelve nuclear weapons tests in the Monte Bello group of islands off the coast of Western Australia, at Emu Field and at Maralinga in the South Australian desert. Prime Minister Menzies agreed to the tests and for using the Australian army as ground support without even putting the proposal to his cabinet for approval. My father Ray Phillipson served in the Australian army at Maralinga. He was scheduled to work on the Buffalo series of four explosions over a three-month period.

As a child, this aspect of my father's army career fascinated me. It seemed other worldly to think that my dad had witnessed nuclear bombs explode not even two miles from where he had been standing. I once asked him what it was like to stand so close. He stated quite plainly that he recalled a flash, a gush of wind and a smell not unlike a NSW Cityrail train when it pulls up the brakes hard into a station. He added that when he went into ground zero to recover cameras and film, the ground crackled underneath his feet. The heat was so intense that in some spots it had turned the sand to glass.

As I grew older, I understood more about the implications of his participation at Maralinga. It wasn't just a boy's own adventure, it was an ongoing risk to his

---

[18]   Grabosky, P N, 'A Toxic Legacy: British Nuclear Weapons Testing in Australia', *Wayward Governance: Illegality and Its Control in the Public Sector*, Canberra, Australian Institute of Criminology. pp. 235–253.

health – one that he was not fully informed about at the time.

It seemed unbelievable to me that the soldiers didn't know of the risks. Hiroshima happened years before the tests. But there they are in the archival footage, smiling and watching in awe as the mushroom clouds climb higher into the sky above their heads. They are dressed only in shorts and shirts. Their ignorance to the danger almost seems surreal.

Even if he had known, my father was not the type to disobey a direct order. He diligently joined the parade in the field, turned his back when instructed, and waited until the explosion went off. At this point, he received heat flashes on the back of his uncovered neck and was asked to turn around. He remembers physically seeing the blast hurtle across the barren landscape towards him.

About ten years after the tests, the veterans who were there started to notice higher than normal rates of cancer as well as infertility, miscarriages, birth defects and mental illness in their children. An association was formed to campaign for access to veteran benefits and compensation. It is a campaign that has yet to bear fruit.

Service at Maralinga was classified as not active duty since no bullets or enemy combat were involved. Those who served there but not elsewhere were not eligible for access to veterans affairs benefits, such as the white or gold card. My father had also served in the Malaya Intervention so his health costs have thankfully been covered.

In 2006, a report[19] commissioned by the Australian government showed that Australians working at the Maralinga and Emu Field testing sites had a twenty-three per cent higher risk of developing cancer than the general population and were eighteen per cent more likely to die from those cancers. Bullets or not, these men have suffered because of their duty.

Over the years, my father testified at a royal commission into the tests and added his name to class actions against the British government seeking compensation for the fatherless families — rare points of protest by him. And now there is the latest appeal to the Human Rights Commission[20]. It is more like an appeal to common decency to right the wrong of how these young men were used and damaged. I doubt much will come of it.

There is seemingly no impetus to provide these men with access to medical care, let alone compensation for what they and their families have suffered because the Australian government of the time decided to use them as guinea pigs.

Another example is from **John Hutton** who became a spokesman for the soldiers who had worked on the site and he has described some of the duties he had to

---

[19]   http://www.dva.gov.au/aboutDVA/publications/health_research/nuclear_
       test/dosimetry/Documents/dosimetry_executive_summary.pdf

[20]   http://www.theaustralian.com.au/news/nation/maralinga-veterans-com
       plain-to-australian-human-rights-commission/story-e6frg6nf-1226582940487

undertake and the operating conditions in a submission to the Department of Veteran Affairs 2008.

I had two tours of duty at Maralinga. The first from 1st February 1956 until 5th September 1956, during this time my posting was to Army Task Force Weapons Trial where most of my time was spent building towers for the forthcoming tests. We made weekly trips to Emu Fields, under orders, as part of a team to bring back spare equipment in the way of tools, cooking utensils, and other items for use at Camp 43, because we were seriously short of supplies. I was not at Maralinga when the first four bombs (Operation Buffalo) were detonated.

I returned on 11th April 1957 and remained until 24th November 1957 (posted to Maralinga Range Support Unit) and was present for the three Antler explosions. I lived at Roadside; a camp situated about eight miles south of the ground zero areas. I was mainly involved in manual labour tasks, building bays A to Z for the scientists to lay out their instruments and other equipment. This involved drilling in star pickets and stringing with fencing wire. We worked in extremely dusty conditions resulting in us returning to our tents every night completely covered in dust, faces included.

Shortly before each bomb was exploded, I and a team of four other engineers would seal the entrance to one of the 'instrument bunkers' with about a thousand sand bags, which we had previously filled. Then about half an hour after each explosion we would return in my Land Rover and remove the sandbags. The bunkers were very close to ground zero and the task took about an hour. We did not wear protective clothing and the bulldust was so heavy that we wore handkerchiefs over our mouths. On return, we were not checked for radiation and had to spend much time in the showers. I was in charge of Sandbag Party B. It was impossible to work wearing a respirator for either sandbagging or driving.

In August 1957 a number of the troop became ill with persistent vomiting but were reluctant to seek medical help for fear of being called 'shirkers'. I was admitted to the Maralinga Village Hospital on 29th August 1957 and did not return to work until after 7th September. I was treated with Largactil which I now know is used not only for vomiting but psychosis and radiation sickness.

In 2001, Dr Sue Rabbit Roff, a researcher from the University of Dundee, uncovered documentary evidence that troops had been ordered to run, walk and crawl across areas contaminated by the Buffalo tests in the days immediately following the detonations, a fact that the British government later admitted. Dr Roff stated that "it puts the lie to the British government's claim that they never used humans for guinea pig-type experiments in nuclear weapons trials in Australia."

There was an unsuccessful legal action undertaken in Britain in 2010, largely comprising of Australian veterans. Sixty years after the first servicemen sailed for the Monte Bello test sites, the UK Supreme Court has narrowly dismissed a class-action lawsuit by nuclear veterans[21] on statute-of-limitation grounds. The court's decision, which could affect hundreds of nuclear veterans, illustrates a legal paradox: the participants in the UK nuclear programme did not sue the government within a required three-year time frame because they did not have the documentation to prove their radiation exposures caused illness, as the British legal system defines causation. They did not have the documentation, apparently because the government had failed to create or safeguard it.

In 2013 Joshua Dale, a Sydney lawyer, lodged a claim on behalf of those veterans with the Australian Human Rights Commission (AHRC). The *Australian Newspaper* released an article on 21st February, 2013:

'Hundreds of Maralinga veterans and their families have today made a complaint to the Australian Human Rights Commission over exposure to nuclear testing in the 1950s and 1960s. The action by 295 veterans is a last-ditch bid to be recognised for having been present when the British government detonated a series of nuclear bombs in remote South Australia.

'Stacks/Goudkamp human rights lawyer Joshua Dale lodged the complaint in Sydney, claiming the Menzies government breached their human rights by ordering their exposure to the harmful effects of radiation in full knowledge of the potential health impacts.

'"This really is the end of the line for the veterans, it's their last chance to make an application and get recognition," Mr Dale said.

'"The government spends so much money defending their access to treatment, we all know they spend a great deal on legal fees – those sorts of funds should be allocated to veterans."

'Mr Dale argued the federal government breached articles three, five and twenty-five of the Universal Declaration of Human Rights, which was the only declaration in place at the time of the nuclear explosions. The breaches relate to their rights to life and liberty; not be tortured or subject to cruel and inhuman treatment; and the right to a standard of living adequate for health and well-being. The veterans involved in the AHRC complaint were also involved in a joint bid for compensation in Britain, which was rejected last year.

'The statute of limitations has prevented the veterans from pursuing their case through Australian courts. About 8,000 veterans were involved in the testing at Maralinga, including mechanics, builders, engineers and army personnel.

'Mr Dale said about 2,000 were still alive. He said those nuclear veterans had been found to have a three times higher rate of cancer than the general population, thirty per cent higher death rates and the rate of abnormality in their children was three times higher than normal. However, any finding made by the AHRC would not be binding and it could not force the federal government to hand over compensation or provide greater rights for veterans.'

---

[21]   http://www.supremecourt.gov.uk/decided-cases/docs/UKSC_2010_0247_ Judgment.pdf

There have been other class actions by Australian ex-armed forces against both the Australian and UK governments but they have not been successful on the grounds that the dosage was not that great in comparison with the Japanese and finally that it all happened a long time ago.

Besides the radiation effects on veterans there was the problem of making the Maralinga site safe from a radiation viewpoint. The initial Maralinga cleanup operation was code named Operation Brumby, and was conducted in 1967. Attempts were made to dilute the concentration of radioactive material by turning over and mixing the surface soil. Additionally, the remains of the firings, including plutonium-contaminated fragments, were buried in 22 concrete-capped pits.

By the 1980s some Australian servicemen and traditional Aboriginal owners of the land were suffering blindness, sores and illnesses such as cancer. They 'started to piece things together, linking their afflictions with their exposure to nuclear testing'. Groups including the Atomic Veterans Association and the Pitjantjatjara Council pressured the Australian government, until in 1985 it agreed to hold a royal commission to investigate the damage that had been caused.

The McClelland Royal Commission into the tests delivered its report in late 1985, and found that significant radiation hazards still existed at many of the Maralinga test sites, particularly at Taranaki, where the Vixen B trials into the effects of burning plutonium had been carried out. A Technical Assessment Group (TAG) was set up to advise on rehabilitation options, and a much more extensive cleanup programme was initiated at the site.

The TAG report plan was approved in 1991 and work commenced on site in 1996 and was completed in 2000 at a cost of $108 million. In the worst-contaminated areas, 350,000 cubic metres of soil and debris were removed from an area of more than two square kilometres, and buried in trenches. Eleven debris pits were also treated with in-situ vitrification. Most of the site (approximately 3,200 square kilometres) is now safe for unrestricted access and approximately 120 square kilometres is considered safe for access but not permanent occupancy. Alan Parkinson[22] has observed that 'an Aboriginal living a semi-traditional lifestyle would receive an effective dose of 5 mSv/a (five times that allowed for a member of the public). Within the 120 km², the effective dose would be up to thirteen times greater.'

One author suggests that the resettlement and denial of Aboriginal access to their homelands 'contributed significantly to the social disintegration which characterises the community to this day. Petrol sniffing, juvenile crime, alcoholism and chronic friction between residents and the South Australian police have become facts of life.' In 1994, the Australian government reached a compensation settlement with Maralinga Tjarutja, which resulted in the payment of $13.5 million in settlement of all claims in relation to the nuclear testing.

Moreover, despite the evidence of the 2006 report (see page 180) on the

[22] *Maralinga: Australia's Nuclear Waste Cover-up*, Alan Parkinson, ABC Books, 2007.

Defence personnel watching blast unprotected at Maralinga.

possible effects of testing, in a blow to the men's claims for compensation, it concluded it was impossible to state whether that was due to the men's exposure to radiation.

Christmas Island

A similar fight has been waged by more than 1,000 UK veterans who took their fight to the Supreme Court – the highest court in the UK – in November 2011 after nearly two years of battles with the MOD in the High Court and the Court of Appeal. In these cases lack of radiation protection was considered at Christmas Island as well as in Australia.

In 2009, ten lead claimants won the first round of the veterans' battle when a High Court judge said claims could go ahead. However the Supreme Court by a verdict of four to three in March 2012 ruled that the events occurred too long ago.

A typical case was that of **Private John Hall**, a 19-year-old RAF ground crew in 1958 whose job it was to wash down contaminated aircraft after they had flown through the radiation clouds. He was given no protective clothing. Instead, he and his fellow RAF servicemen had been ordered to turn away from the mushroom cloud and put their hands in front of their faces. He later said that as he did so, his hands 'lit up like an X-ray', and he saw his bones outlined through the flesh.

His part was small, but he knew that it was vital work. If open warfare with the USSR was to be avoided, it was imperative that Britain should develop a nuclear deterrent. But as he helped to end the Cold War in the dust and blistering heat, John had no idea that he and many of his fellow servicemen would later suffer ill health and premature death. His own would come at the age of 53, after a long struggle against hairy cell leukaemia, a rare form of cancer that affects just 200 people per year in Britain. He would not have suspected that his children, like those of hundreds of nuclear veterans, would be born with congenital deformities and unidentified illnesses.

Scientists believe that exposure to radiation can cause genetic damage, resulting in the development of new hereditary disorders; Mr Hall's eldest son Colin, now 50, has deformed hands, and his other son Ian, 44, was born without calf muscles. Other children of nuclear veterans have suffered chronic musculoskeletal disorders; deformity of the hands, feet, bladder and genitals; heart malformation; hearing defects; spinal bifida, and a host of other illnesses. Many have decided not to have children, for fear of perpetuating genetic abnormalities.

It has also been reported that there has been a disproportionate number of stillbirths and fatal deformities among descendants of nuclear veterans. One recently gave birth to a baby with its head attached to one shoulder; it survived for less than two days.

A further study undertaken by Dr Sue Rabbitt Roff in 1999, found that of 2,261 children born to veterans, 39 per cent were born with serious medical conditions. By contrast, the national incidence figure in Britain is around 2.5 per cent. In 1958, Mr Hall could not have known about any of

Private John Hall who decontaminated Valiants after they had dropped bombs.
(*The Daily Telegraph*)

this. And he would never have dreamt that the British government – unlike those of other Western countries – would consistently refuse to give the 3,000 nuclear veterans and their families any compensation, or even any special recognition.

Indeed, the MOD has spent more than £4 million blocking legal claims brought by hundreds of nuclear veterans and their families. A dedicated charity, the British Nuclear Test Veterans' Association (BNTVA), has been campaigning on the issue for decades. Earlier this year, a letter addressed to John Baron MP, the patron of the BNTVA, from David Cameron, the prime minister, seen by *The Sunday Telegraph*, made clear that the Government's position had not changed:

'The Government and I continue to recognise and be grateful to all servicemen who participated in the British nuclear testing programme, but it would be divisive to offer nuclear test veterans this level of recognition for being involved in the project, when those who have undertaken other specialist duties would not be receiving the same. I can therefore only reiterate that I will not be making an oral statement on this subject to the House.'

There are many examples of people who witnessed the UK tests giving details of the effects on them but to tell all their stories would be outside the scope of this book. The aim of this appendix is to try to ensure that the apparent downside of the successful tests are not ignored.

WISLEY LOG BOOK – SOME PERSONAL MEMORIES

1956 Month	Date	AIRCRAFT Type	No.	Pilot, or 1st Pilot	2nd Pilot, Pupil or Passenger	DUTY (Including Results and Remarks)	DAY (1)
—	—	—	—	—	—	Totals Brought Forward	92·30
APRIL	5	CHIPMUNK T10	WD373	SELF	F/O OFF ALLEN	CHECK	
APRIL	5	CHIPMUNK T10	WD 373	SELF	CADET F'HAOGH	C.C.F. AIR EXPERIENCE	
APRIL	5	CHIPMUNK T10	WD 373	SELF	CADET FLETCHER	C.C.F. AIR EXPERIENCE	
		VICKERS ARMSTRONG LTD, WISLEY.					
APRIL	10	VALIANT B1	WZ 394	MR. P. MURPHY	SELF & CREW	POST RECTIFICATION CHECK	
APRIL	12	VALIANT B1	WZ 396	MR. J. JARVIS	SELF & CREW	AIRTEST – FLAP OPERATION	
APRIL	14	VALIANT B1	WB 215	MR. P. MURPHY	SELF & CREW	WISLEY – BOSCOMBE DOWN	
APRIL	19	VALIANT B1	WZ 397	MR. J. JARVIS	SELF & CREW	PRODUCTION AIR TEST (½ × F/T A)	
APRIL	20	VALIANT B1	WP 210	MR. E.B. TRUBSHAW	SELF & CREW	CHECK ON MALFUNCTIONING D-DOORS & O/W TANKS	
APRIL	21	VALIANT B1	WZ 398	MR. S. HARRIS	SELF & CREW	1ST FLIGHT – BROOKLANDS TO WISLEY	
APRIL	25	VALIANT B1	WZ 375	MR J. JARVIS	SELF & CREW	PRODUCTION AIR TEST	
APRIL	27	VALIANT B1	WZ 375	MR B.G. ASTON	SELF & CREW	POST RECTIFICATION AIR TEST	
APRIL	28	VALIANT B1	WZ 398	MR W.D. JARVIS	SELF & CREW	AIR TEST (FINAL PRODUCTION)	
APRIL	28	VALIANT B1	WZ 399	MR B.G. ASTON	SELF & CREW	BROOKLANDS – WISLEY (1ST FLIGHT)	
APRIL	30	VALIANT B1	WZ 375	MR B.G. ASTON	SELF & CREW	WISLEY – BOSCOMBE DOWN	

E.B. Trubshaw.

E.B. TRUBSHAW

DEPUTY CHIEF TEST PILOT

VICKERS ARMSTRONG (A/C) LTD.

SUMMARY FOR APRIL 1956 UNIT DATE 1ST MAY 1956 SIGNATURE R. Hayward

T Y P E S { 1. CHIPMUNK T10 2. VALIANT B1 }

GRAND TOTAL [Cols. (1) to (10)] 1024 Hrs. 55 Mins	Totals Carried Forward	92·30 (1)

1956 March	Date	AIRCRAFT Type	No.	Pilot, or 1st Pilot	2nd Pilot, Pupil or Passenger	DUTY (Including Results and Remarks)	
—	—	—	—	—	—	Totals Brought Forward	
		E.B. Trubshaw		SUMMARY FOR JUNE 1956		T Y P E S { 1. VALIANT B1 2. 3. }	
		DEPUTY CHIEF TEST PILOT		UNIT			
		VICKERS ARMSTRONG (A/C) LTD.		DATE 2ND JULY 1956			
				SIGNATURE R. Hayward F/O OFF.			
JULY	2	VALIANT B1	WZ 402	MR. D.G. ADDICOTT	SELF & 2 CREW	AIR TEST – POST RECTIFICATION CHECK.	
JULY	3	VALIANT B1	WP 208	MR. T.S. HARRIS	SELF & 2 CREW	AUTO STAB, AUTO PLT, & S/G. AT 5×, 20×30×%	
JULY	4	VALIANT B1	WP 208	MR. T.S. HARRIS	SELF & 3 CREW	AUTOMATIC I.L.S. APPROACHES AT WITTERING.	
JULY	9	VALIANT B1	WZ 376	MR. B.G. ASTON	SELF & 2 CREW	HOSE DROGUE UNIT AIRFLOW TEST AT 200 KT	
JULY	10	VALIANT B1	WP 208	MR. T.S. HARRIS	SELF & 3 CREW	AUTOMATIC I.L.S. APPROACHES · WITTERING.	
JULY	10	VALIANT B1	WZ 376	MR. B.G. ASTON	SELF & 2 CREW	HOSE DROGUE UNIT AIRFLOW TEST 250 KT AT 10"	
JULY	11	VALIANT B1	WZ 376	MR. B.G. ASTON	SELF & 3 CREW	HOSE DROGUE UNIT AIRFLOW TEST 300 KT AT 10 FT.	
JULY	11	VALIANT B1	WZ 390	MR. D.G. ADDICOTT	SELF & 3 CREW	TEST ON VARIABLE REAR STAGE ENGINES.	
JULY	12	VALIANT B1	WZ 390	MR. D.G. ADDICOTT	SELF & 3 CREW	PERFORMANCE AIR TEST – CRUISE CLMB, DRICING	
JULY	17	VALIANT B1	WZ 405	MR. E.B. TRUBSHAW	SELF & 2 CREW	WISLEY – MARHAM – DELIVERY	
JULY	20	VALIANT B1	WZ 390	MR. T.S. HARRIS	SELF & 3 CREW	OBSERVATION POST FOR WZ 376 TANKER	
JULY	23	VALIANT B1	WP 204	MR. T.S. HARRIS	SELF & 3 CREW	GENERATOR AIR TEST AT MAX LOAD 22½×30	200×

E.B. Trubshaw.

DEPUTY CHIEF TEST PILOT

VICKERS ARMSTRONG (A/C) LTD.

SUMMARY FOR JULY 1956 UNIT DATE 27TH JULY 1956 SIGNATURE R. Hayward F/O OFF.

T Y P E S { 1. VALIANT B1 2. 3. } TOTAL VALIANT :–

GRAND TOTAL [Cols. (1) to (10)] 1060 Hrs. 45 Mins	Totals Carried Forward	

Dick Hayward at Wisley.

BUFFALO LOG BOOK –
AN IMPORTANT RECORD

Page from Ted Flavell's log book. (*Roger Flavell*)

Interestingly Ted Flavell did not make any special mark on 11th October when he dropped the first UK atom bomb. John Ledger thought it was worth recording, as can be seen below.

Extract from John Ledger's log book. (*Andrew Arnold*)

ACRONYMS AND EXPLANATIONS

A&AEE	Aircraft and Armament Experimental Establishment
AAR	Air-to-air refuelling
AEO	Air Electronics Officer
AFB	United States Air Force Base
AHRC	Australian Human Rights Commission
AOC	Air Officer Commanding Group
ASF	Aircraft Servicing Flight
ATC	Air Traffic Control
AVS	Air Ventilated Suits
AWR	Air Weapons Range
BCAS	Bomber Command Armament School
BNTVA	British Nuclear Test Veterans' Association
BoI	Board of Inquiry
BTR	Bomber Training Requirements
CND	Campaign for Nuclear Disarmament
ECM	Electronic Counter Measures
EEF	Electronic Engineering Flight
ETPS	Empire Test Pilots School
F700	Servicing record book for each aircraft
GPI4	Ground Position Indicator
'Green Satin'	Doppler drift and ground speed computer
H2S	Early radar fitted to Victors and Vulcans
HDU	Hose Drum Unit
HE	High Explosive
ICC	Intermediate Co-pilots Course
ILS	Instrument Landing System
INS	Inertial Navigator System
IRE	Instrument Rating Examiner
JARIC	Joint Air Reconnaissance Intelligence Centre

JNCO	Junior Non-Commissioned Officer
MC	Medium Capacity
MOD	Ministry of Defence
mSv	Millisieverts
MT	Mechanical Transport
NATO	North Atlantic Treaty Organisation
NBS	Navigation and Bombing System
OCU	Operational Conversion Unit
ORP	Operational Readiness Platform
QFI	Qualified Flying Instructor
QRA	Quick Reaction Alert
RAE	Royal Aircraft Establishment
RATOG	Rocket-Assisted Take-Off Gear
RBSU	Radar Bomb Score Unit
RCAF	Royal Canadian Air Force
SAC	Strategic Air Command
SACEUR	Supreme Allied Commander Europe
SAM	Surface-to-Air Missiles
SHQ	Station Headquarters
SNCO	Senior Non-Commissioned Officer
SOP	Standard Operating Procedure
STC	Strike Command
Sunspot	Detachment to Malta
TACAN	Tactical Air Navigation System
TFR	Terrain Following Radar
USAF	United States Air Force
WRE	Weapons Research Establishment, Woomera, Australia

INDEX

COMPLETE TONY BLACKMAN'S
TRILOGY OF THE THREE V BOMBERS

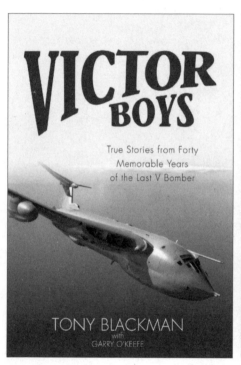

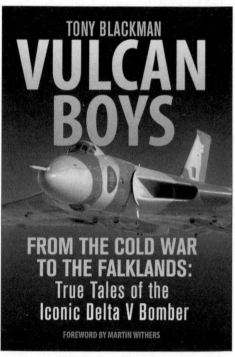

VICTOR BOYS AND VULCAN BOYS
AVAILABLE NOW